Energy Geostructures

Energy Geostructures

Innovation in Underground Engineering

Edited by
Lyesse Laloui
Alice Di Donna

Series Editor
Gilles Pijaudier-Cabot

ISTE

WILEY

First published 2013 in Great Britain and the United States by ISTE Ltd and John Wiley & Sons, Inc.

ISTE Ltd
27-37 St George's Road
London SW19 4EU
UK

www.iste.co.uk

John Wiley & Sons, Inc.
111 River Street
Hoboken, NJ 07030
USA

www.wiley.com

Library of Congress Control Number: 2013941765

British Library Cataloguing-in-Publication Data
A CIP record for this book is available from the British Library
ISBN: 978-1-84821-572-6

MIX
Paper from responsible sources
FSC® C013604

Printed and bound in Great Britain by CPI Group (UK) Ltd., Croydon, Surrey CR0 4YY

Table of Contents

Preface

Energy geostructures are spreading rapidly all around the world. They represent a renewable and clean source of energy that can be used for the heating and cooling of buildings, infrastructures and all kinds of environments. This technology links the structural role of geostructures with the energy supply, using the principle of shallow geothermal energy. On the one hand, heat is extracted from the ground during winter in order to satisfy heating needs and, on the other hand, heat is injected into the ground during summer to satisfy cooling needs.

The dual role of these structures makes their design challenging and more complex with respect to conventional projects. Besides the known procedures that are currently applied for the delivery of a geostructure, a number of additional issues arise when also deciding to exploit it from the energy supply point of view. These include the design and dimensioning of the geothermal equipment, study of the energy demand and the consequent optimization of the system, consideration of additional effects induced by the temperature variation on the structure itself in terms of stresses and displacements and the definition of the responsibilities of different professionals involved in the project. As energy geostructures are a new engineering technology, there is a need for improved scientific knowledge and defined design procedures.

The objective of this book is to provide the readers with an exhaustive overview of the up-to-date knowledge available on these structures as well as the procedures currently applied in the regions where they are implemented. The book is divided into four parts, each of them divided into chapters written by the brightest engineers and researchers in the field. Part 1 deals with the physical modeling of energy geostructures, including laboratory investigation of the thermomechanical behavior of soils, *in situ* analyses, centrifuge testing and small-scale experiments. Part 2 includes numerical simulation results of energy piles, tunnels and bridge foundations, while also considering the implementation of such structures in different climatic areas. Part 3 discusses the practical engineering aspects, from the delivery of energy geostructures to the development of design tools for their geotechnical dimensioning. Finally, the book concludes with a real case study.

The editors would like to thank all the authors for their innovative contributions.

Lyesse LALOUI and Alice DI DONNA
July 2013

PART 1

Physical Modeling of Energy Piles at Different Scales

Chapter 1

Soil Response under Thermomechanical Conditions Imposed by Energy Geostructures

The foundation of a building represents a connection between the structure and the supporting soil. The mechanical loads coming from the structure are transferred to the soil through it. A number of requirements must be fulfilled to ensure the stability and comfort of the over-structure, the most important of which are (1) the admissible displacements, (2) the acceptable (concrete) stresses and (3) the safety margins with respect to failure [BSI 95]. These aspects are related to the types and properties of the surrounding soils. Data concerning the soil's response must be collected through a geotechnical survey and represent the basis for the design of the required foundations. Therefore, the behavior of the soil plays a primary role in the design of every geostructure, i.e. every structure that transfers a load to the ground. In the case of energy geostructures, an energy supply role is added to the conventional role of the foundation as a structural support. The foundation is thus subjected to both mechanical and thermal loads transmitted from the piles to the ground. This is the main motivation for understanding and modeling the soil's response when subjected to a thermomechanical solicitation. In this chapter, the state of knowledge on the thermomechanical behavior of soils is revised within the framework of energy geostructures. A constitutive model capable of reproducing the described behavior is presented and used to study the response of soils subjected to thermal-stress paths typical of the areas around energy piles.

Chapter written by Alice DI DONNA and Lyesse LALOUI.

1.1. Introduction

Deep foundations are usually used to limit the settlements of buildings, increase capacity with respect to shallow foundations or reach a more resistant layer of soil when the quality of the surface soil is low. Two stages of the geotechnical design of such foundations are related to the behavior of the surrounding soil: the evaluation of the geotechnical bearing capacity and the prediction of displacements. Starting from the equilibrium of a pile (Figure 1.1), the maximum load Q_{LIM} that a pile can support is calculated as:

$$Q_{LIM} = Q_S + Q_P - W_P \qquad [1.1]$$

where Q_S is the portion of bearing capacity provided by the friction between the pile and the soil, Q_P is the portion of the bearing capacity provided by the soil below the pile tip and W_P is the pile weight [LAN 99]. A general formula for the calculation of the lateral and base components is:

$$Q_S = \int_0^{\mathrm{H}} \int_0^{2\pi} \sigma'_h \cdot \tan(\delta) \cdot \mathrm{R}\, d\omega dz \qquad [1.2]$$

$$Q_P = \int_0^{R} \int_0^{2\pi} (9 \cdot \mathrm{Cu} + \sigma_v) \cdot \mathrm{r}\, d\omega dr \qquad [1.3]$$

where H is the pile height, σ'_h is the horizontal effective stress normal to the pile–soil interface, δ is the friction angle at the interface, R is the pile radius, Cu is the undrained shear strength, σ_v is the vertical stress at the pile tip and r, ω and z are the radial, circumferential and vertical cylindrical coordinates, respectively. From these equations, it appears that the lateral resistance depends, apart from the friction angle at the interface, on the stress state at the pile–soil interface, while the tip resistance is directly related to the resistance of the soil below the pile.

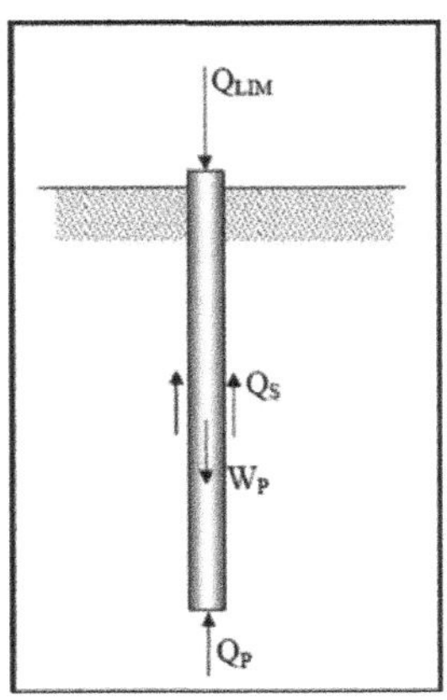

Figure 1.1. *Equilibrium of a pile*

When a thermal load is transmitted from a pile to the soil, the latter reacts by changing its volume, and eventually its response, depending on the type of soil. As a result, the temperature variation can affect (positively or negatively) the stress state at the pile–soil interface and the shear strength that governs tip resistance. More significantly, the thermal volume variation in the soil affects the foundation displacement, making it move upward when the soil dilates and downward when it contracts. The entity of the effects induced by the temperature variations on the behavior of the foundation depends on the volume of heated ground and the range of temperature variation. The thermal load imposed by energy piles in current applications is in the range of 2–30°C (see Chapter 3 for data related to current applications), but future developments using the injection of heat in the ground from other technologies, such as solar panels, will lead to higher temperature variations.

1.2. Thermomechanical behavior of soils

Soils are porous materials made up of a solid skeleton, represented by grains or aggregates, and pores that, in saturated conditions, are filled with water. The case of partial saturation is not considered in this chapter (see Chapter 8 for more details concerning energy piles in unsaturated soils and [FRA 08] for the non-isothermal behavior of unsaturated soils). Soils can be divided into two families: (1) granular (sand and gravel) and (2) fine-grained (silt and clay) materials. Heating a sandy soil in drained conditions results in an increase of volume directly related to the grains' thermal expansion coefficient. Also, water dilates thermal elastically with a thermal expansion coefficient usually higher than that of grains, but due to drained conditions, the water is free to flow away and does not contribute to the volume variation of the material itself. Table 1.1 provides the thermal expansion coefficients of some minerals and water.

Material	**Volumetric thermal expansion coefficient [10^{-6}°C^{-1}]**
Muscovite	24.8
Kaolinite	29.0
Chlorite	31.2
Illite	25.0
Smectite	39.0
Water	139 + 6.1·T

Table 1.1. *Volumetric thermal expansion coefficients (T stands for temperature) [MCK 65, DIX 93]*

The response of clays is more complicated and will be discussed in the next section. The complexity of clayey materials' thermomechanical behavior is a direct consequence of their microstructure and the electrochemical equilibrium between clay particles. Details about this aspect can be found, among others, in [HUE 92].

1.2.1. *Thermomechanical behavior of clays*

As for granular materials, the two constituents of saturated fine-grained materials (grains/aggregates and water) undergo thermoelastic expansion when heated. However, it has been proved through experimental testing that either a contractive or a dilative volume variation can be observed during heating in drained conditions depending on the load history. The latter is commonly described through the overconsolidation ratio (OCR), defined as:

$$\mathrm{OCR} = \frac{\sigma'_p}{\sigma'_v} \qquad [1.4]$$

where σ'_p is the preconsolidation stress and σ'_v is the current vertical effective stress. The preconsolidation stress is the maximum vertical stress that the soil has already supported (load history). The soil retains a memory of the maximum charge that it has already supported, so that if it is subjected to a load lower than the preconsolidation stress, its deformation is relatively small and, above all, reversible (elastic). If the applied load reaches and surpasses the initial preconsolidation stress, the deformation becomes more significant and, above all, partially irreversible (elastoplastic). In this sense, the preconsolidation stress corresponds to the maximum experienced density (or the lowest void ratio). From a mechanical point of view, it is used to be defined as the limit between the elastic and elastoplastic domains in terms of applied stresses. A soil is considered normally consolidated (NC) if the OCR is within the range of 1 and 2; i.e., if the current load is close to the maximum that the soil has ever supported. Conversely, the material is said to be overconsolidated (OC) if the OCR is greater than 2; i.e., if the current load is lower than the historical maximum. In terms of a fine-grained soil's response to a temperature variation in drained conditions, it has been largely demonstrated that the material contracts upon heating in NC conditions and a significant part of this deformation is irreversible, while highly OC materials experience a volume expansion during heating that is recovered during cooling. Between these two extreme cases, there is an intermediate case represented by slightly OC clays. In this case, the material shows initial dilation and subsequent contraction during heating, followed by contraction during cooling, thus representing a transition between the two main cases. The first experimental results of this nature date back to between the 1960s and 1980s [CAM 68, PLU 69, DEM 82, DES 88, BAL 88] and have been widely confirmed more recently [MIL 92, TOW 93, BUR 00, CEK 04]. Similar results have been obtained by the authors for a wide range of different clayey materials containing variable quantities of illite, kaolinite, chlorite and smectite. Some examples are given in Figure 1.2. In other words, these experimental results show that a soil can undergo irreversible deformation due to an increase in temperature under a constant mechanical load equal to (see NC cases in Figure 1.2) or even slightly lower than the preconsolidation stress (see cases with OCR = 2 in Figure 1.2). A number of experimental studies on various clays have been performed to develop a theoretical framework for describing this phenomenon, and have led to the conclusion that the “apparent” preconsolidation stress decreases at constant void ratio with increasing temperature. The word “apparent” is used to underline the fact that the applied mechanical load does not change, so that the maximum load historically applied is always the same. Some of these results are summarized and compared in Figure 1.3.

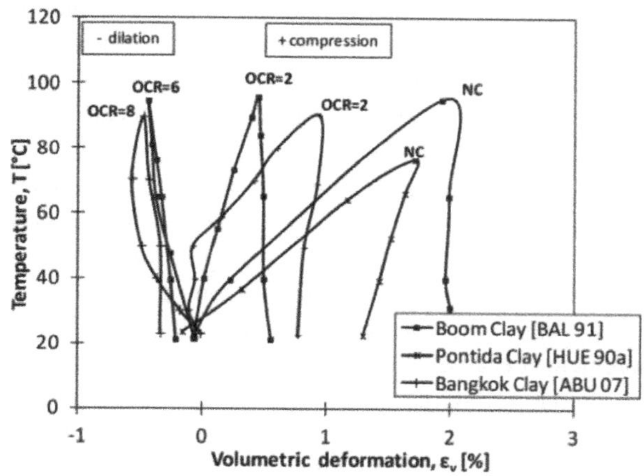

Figure 1.2. *Thermal deformation of various clays under different initial conditions*

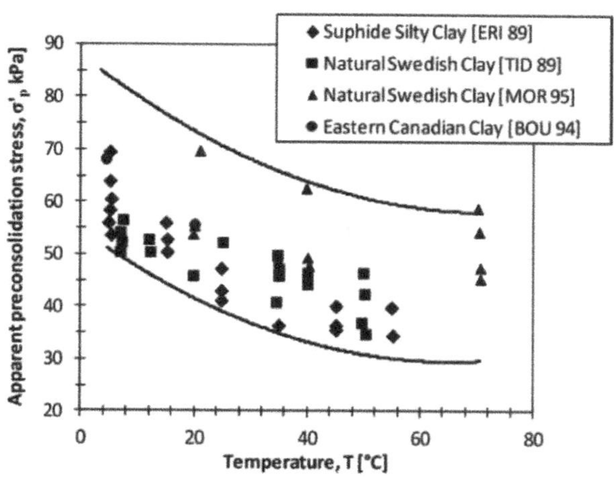

Figure 1.3. *Influence of temperature on preconsolidation pressure [LAL 03]*

To fit these results within a theoretical-schematized framework, the evolution of the apparent preconsolidation pressure can be plotted in the mean effective stress-temperature plane, as shown in Figure 1.4, where the mean effective stress p′ is defined as:

$$p' = \frac{\sigma'_v + 2\cdot\sigma'_h}{3} \qquad [1.5]$$

The isotropic preconsolidation pressure, or the maximum mean effective stress that the soil has ever supported, is considered in this plane. As discussed earlier, and in light of the results illustrated so far, the (apparent) preconsolidation pressure represents the limit between the elastic and elastoplastic domains. In Figure 1.4, point A represents the state of a material subjected to an initial temperature (T_0) and an OC stress state, as its current mechanical load (p'_A) is lower than the preconsolidation pressure (p'_{prec}). If this material is subjected to drained heating under a constant mean effective stress p'_A, it will move from A to A'. This thermal-stress path remains inside the elastic domain and the material therefore reacts elastically (cases with OCR = 6 or 8 in Figure 1.2). However, considering an NC (point B) or slightly OC (point C) material, heating induces thermoplastic strain as the thermal-stress paths (i.e. B-B' or C-C') encounter the border of the elastic domain. In the NC case, the material responds elastoplastically from the first temperature increment because its initial stress state is already on the border between the two domains (NC cases in Figure 1.2). In the slightly OC case, the response of the material is initially elastic (up to C" in Figure 1.4) and then elastoplastic (cases with OCR = 2 in Figure 1.2). The reduction in the apparent preconsolidation pressure, or that of the elastic domain, with temperature is known as thermal softening.

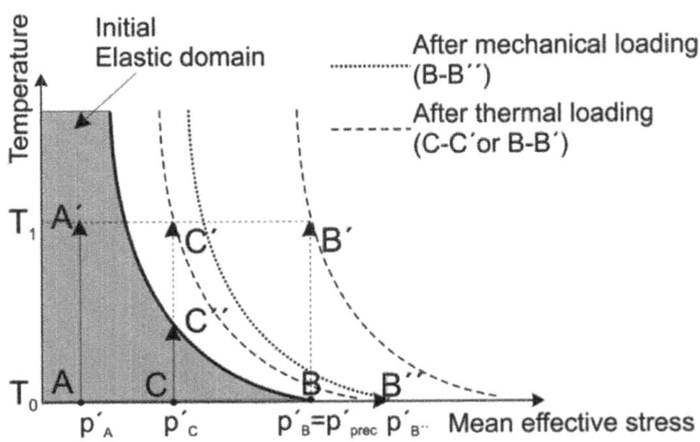

Figure 1.4. *Thermal-stress paths in the mean effective stress-temperature plane*

This phenomenon is opposite of the so-called strain hardening, or the increase of the elastic domain due to the development of irreversible deformations. This latter phenomenon can be explained by simply going back to the definition of preconsolidation stress. If an NC material (point B in Figure 1.4) is mechanically loaded (stress path B-B"), the current applied stress becomes the new preconsolidation stress (the maximum historically applied stress). The increase of the preconsolidation pressure from p'_B to $p'_{B''}$ corresponds to an increase in the elastic domain. The same phenomenon occurs for thermal loading, so that the thermal paths that induce plasticity (C"-C' and B-B' in Figure 1.4) also induce an increase in the elastic domain (from the continuous line to the dotted lines in Figure 1.4).

When non-isotropic stress states are considered, the elastic domain can be represented in a tridimensional space by adding an axis to the plane representation illustrated in Figure 1.4. This third axis corresponds to the deviatoric stress (invariant) q, defined as:

$$q = \sqrt{\frac{3}{2}}\sqrt{\mathrm{tr}(\mathrm{d}\mathbf{s})^2} \qquad [1.6]$$

where tr represents the trace of the tensor and **s** the deviatoric stress tensor. The thermal softening is also represented in this space by the shrinkage of the elastic domain with increasing temperature (Figure 1.5). The main consequence of thermal softening under deviatoric stress states is that if an OC material (with a certain void ratio) is sheared at a high temperature (stress path A′-A″), it reaches the border between the elastic and the elastoplastic domains at a lower deviatoric stress with respect to shearing at the initial temperature (stress path A-A‴).

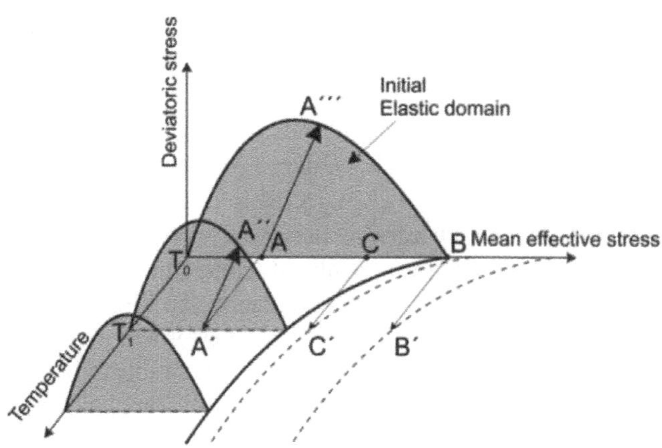

Figure 1.5. *Thermal-stress paths in the mean effective stress–deviatoric stress-temperature space*

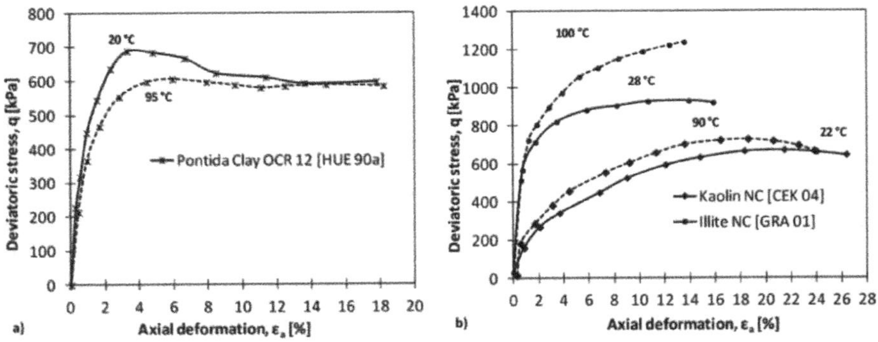

Figure 1.6. *Shear strength at different constant temperatures of a) OC and b) NC clays*

In other words, the material undergoes plasticity earlier than in the isothermal case. Conversely, in the case of an NC or slightly OC material, heating results in a combination of thermal softening and strain hardening. An in-depth analysis of this phenomenon can be found in [HUE 09], and Figure 1.6 illustrates some of those aspects. Within the framework of energy geostructures, the behavior of soils during seasonally cyclic thermal loading is as important as that of the response during a single monotonic heating. Campanella and Mitchell [CAM 68] and Hueckel and Baldi [HUE 98] have shown that the first temperature cycle removes most of the irreversible volume change in NC clays and that subsequent cycles of the same magnitude and range produce small increments of irreversible deformation that decrease cycle after cycle, revealing an accommodation phenomenon. This result is shown in Figure 1.7(a) for illite [CAM 68] and Figure 1.7(b) for a carbonate clay [HEU 98]. Thus, when remaining in the same theoretical framework (Figures 1.4 and 1.5), it is expected that the initial OCR also has an effect on the shear strength of the material at ambient temperature after the application of one or more thermal cycles. If the material is initially OC, a heating-cooling cycle (stress path A-A'-A in Figure 1.4) does not produce any plastic deformation, so that the dimension of the elastic domain after the entire cycle, and thus the response under shearing, is not affected because no permanent change is induced on the void ratio. Conversely, if an initially NC or slightly OC material is subjected to a heating-cooling cycle (B-B′-B or C-C′-C in Figure 1.4), strain hardening occurs as plastic deformation is produced. The material ends up being OC (at an ambient temperature T_0), and this process is generally called thermally induced overconsolidation. To illustrate this phenomenon experimentally, Abuel-Naga *et al.* [ABU 06] has performed a series of tests on an NC soft clay. The result is shown in Figure 1.8(a), which illustrates the effect of one thermal cycle on the oedometric consolidation curve of Bangkok clay. In this test, a conventional oedometric consolidation was run up to 100 kPa at an initial temperature of 22°C (from point 1 to 2). Next, the sample was heated to 90°C and cooled back to the initial temperature (from point 2 to 3). Finally, the consolidation was continued up to 200 kPa (from point 3 to 5). The results plotted in the figure confirm that the thermal cycle induced a thermal overconsolidation: when mechanical consolidation restarted after the thermal cycle (point 3), a higher stress was needed to plastify again (from 3 to 4). Coherently with the results presented in Figure 1.2, the volume decreased during the thermal cycle (from 2 to 3). As expected, the thermally induced consolidation also resulted in an increase in shear strength. An experimental example of this behavior has been provided by Burghignoli *et al.* [BUR 00], who have showed that if an NC sample is heated and then cooled in drained conditions and tested in a triaxial apparatus, its undrained shear strength is higher than that of an equivalent sample tested at constant ambient temperature (Figure 1.8(b)). At the same time, for instance, an increase in shear strength due to thermal consolidation would improve tip resistance.

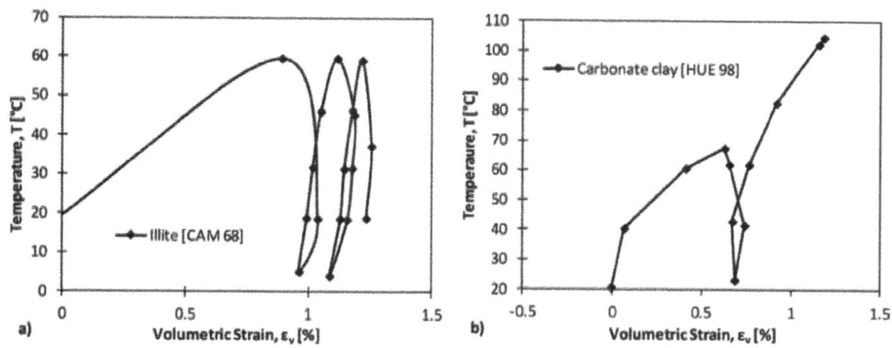

Figure 1.7. *Thermal cyclic effects on NC clays*

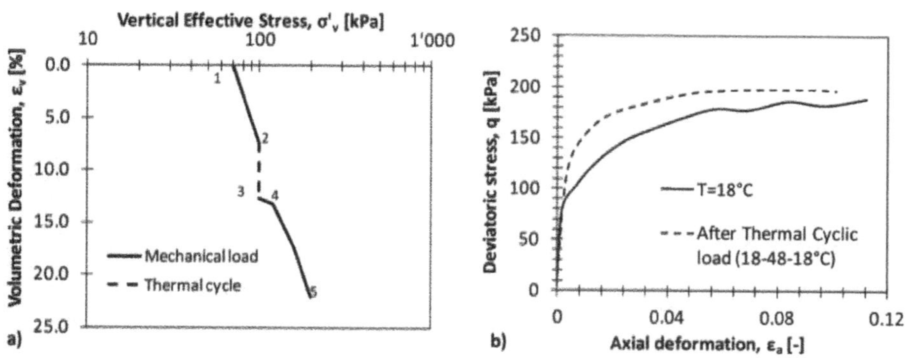

Figure 1.8. *Thermal cyclic effects on a) consolidation of NC Bangkok clay [ABU 06] and b) undrained shear resistance of NC Tody clay [BUR 00]*

1.3. Constitutive modeling of the thermomechanical behavior of soils

This section provides a soil constitutive model with the above-mentioned features and some numerical results. Several constitutive models for describing the thermomechanical behavior of soils have been proposed in the last two decades. A comprehensive summary of their features, capabilities and limitations is presented by Hong *et al.* [HON 13]. Among these models, the model proposed by Laloui *et al.* [LAL 09] is described here.

1.3.1. *The ACMEG-T model*

The Advanced Constitutive Model For Environmental Geomechanics – Thermal effects (ACMEG-T) belongs to the Cam-clay family and is based on the critical state theory. The isothermal part is based on the works of Hujeux [HUJ 79]. Various successive improvements [LAL 93, MOD 97, LAL 08, LAL 09, DI 13] have been made to extend the model's use to non-isothermal conditions. According to the elastoplasticity theory, the tensor of the total strain increment $\mathbf{d\varepsilon}$ is divided into elastic $\mathbf{d\varepsilon^e}$ and plastic $\mathbf{d\varepsilon^p}$ components, so that:

$$\mathbf{d\varepsilon} = \mathbf{d\varepsilon^e} + \mathbf{d\varepsilon^p} \qquad [1.7]$$

The increment of total deformation is composed of volumetric $\mathrm{d}\varepsilon_\mathrm{v}$ and deviatoric $\mathrm{d}\varepsilon_\mathrm{d}$ parts, so that:

$$\mathbf{d\varepsilon} = \frac{\mathrm{d}\varepsilon_\mathrm{v}}{3}\mathbf{I} + \mathbf{de} \quad \text{with} \quad \mathrm{d}\varepsilon_\mathrm{v} = \mathrm{tr}(\mathbf{d\varepsilon}) \quad \text{and} \quad \mathrm{d}\varepsilon_\mathrm{d} = \frac{\sqrt{6}}{3}\sqrt{\mathrm{tr}(\mathbf{de})^2} \qquad [1.8]$$

where $\mathbf{I}$ represents the unitary tensor and $\mathbf{de}$ represents the deviatoric strain increment tensor. The Terzaghi formulation for the effective stresses is introduced, so that:

$$\boldsymbol{\sigma}' = \boldsymbol{\sigma} - \mathrm{p_w}\mathbf{I} \qquad [1.9]$$

where $\boldsymbol{\sigma}'$and $\boldsymbol{\sigma}$ are the effective and total stress tensors, respectively, and $\mathrm{p_w}$ is the pore water pressure. As already mentioned (equations [1.5] and [1.6]), the effective stress increment tensor can be split into the mean effective stress increment dp' and the deviatoric stress increment dq (invariant):

$$\mathbf{d\sigma}' = \mathrm{dp}'\mathbf{I} + \mathbf{ds} \qquad [1.10]$$

where $\mathbf{ds}$ is the deviatoric stress increment tensor. In the elastic non-isothermal domain, the increments of volumetric and deviatoric deformation are equal to:

$$\mathrm{d}\varepsilon_\mathrm{v}^\mathrm{e} = \frac{\mathrm{dp}'}{\mathrm{K_s}} - \beta'_\mathrm{s}\mathrm{dT} \quad \text{and} \quad \mathrm{d}\varepsilon_\mathrm{d}^\mathrm{e} = \frac{\mathrm{dq}}{3\mathrm{G_s}} \qquad [1.11]$$

where K_s is the bulk modulus, β'_s is the volumetric thermal expansion coefficient, T is the temperature and G_s is the shear modulus. The nonlinear elastic response is obtained through the following equations:

$$K_s = K_{ref}\left(\frac{p'}{p'_{ref}}\right)^{n_e} \text{ and } G_s = G_{ref}\left(\frac{p'}{p'_{ref}}\right)^{n_e} \quad [1.12]$$

where K_{ref} and G_{ref} are the two moduli at the reference mean effective stress p'_{ref} and n_e is a material parameter. The plastic response is described by two mechanisms, one isotropic and one deviatoric, that are coupled. Purely isotropic loading causes only volumetric plastic deformation, while purely deviatoric loading causes both deviatoric and volumetric plastic deformation. The two yield surfaces are represented in the p′-q-T space at constant void ratio in Figure 1.9(a). Both are temperature dependent, so that when a thermomechanical load leads the stress point on one of these two surfaces, thermal plastic deformation (contraction) is developed. The isotropic yield limit is:

$$f_{iso} = p' - p'_c r_{iso} = 0 \quad [1.13]$$

where r_{iso} is the isotropic mechanism's degree of mobilization of plasticity (bounding surface theory [DAF 80]). The yield limit evolves with the volumetric plastic deformation induced by the isotropic mechanism $\varepsilon_v^{p,iso}$, as:

$$r_{iso} = r_{iso}^{e} + \frac{\varepsilon_v^{p,iso}}{c+\varepsilon_v^{p,iso}} \quad [1.14]$$

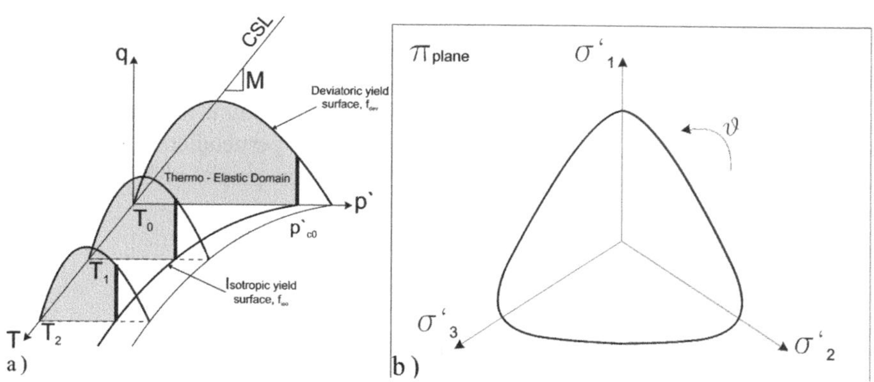

Figure 1.9. *ACMEG-T model (T2>T1>T0): a) representation of the isotropic and deviatoric yield limits as functions of temperature and b) van Eekelen surface and plane strain stress path in the deviatoric plane*

where r_{iso}^{e} is the initial value of the degree of mobilization of plasticity and c is a material parameter. In this model, the dependence of the preconsolidation pressure on temperature (thermal softening) and its evolution with the development of the volumetric plastic deformation (strain hardening) is expressed as [LAL 03]:

$$p'_c = p'_{c0} e^{\beta \varepsilon_v^p} \left[1 - \gamma_T \cdot \ln\left(\frac{T}{T_0}\right)\right] \quad [1.15]$$

where p'_{c0} and T_0 are the initial preconsolidation pressure and temperature, respectively, β is the plastic index, ε_v^p is the total volumetric plastic deformation and γ_T is a material parameter that defines the shape of the isotropic yield function with respect to temperature (horizontal plane in Figure 1.9(a) or Figure 1.4). The deviatoric yield limit is:

$$f_{dev} = q - Mp'\left(1 - b \cdot \ln\left(\frac{d \cdot p'}{p'_c}\right)\right) r_{dev} = 0 \quad [1.16]$$

where M is the slope of the critical state line (CSL) in the p′-q plane (Figure 1.9(a)), b and d are two material parameters and r_{dev} is the deviatoric mechanism's degree of mobilization of plasticity. The latter has the same role as r_{iso} and is given by:

$$r_{dev} = r_{dev}^e + \frac{\varepsilon_d^p}{a + \varepsilon_d^p} \quad [1.17]$$

where a is a material parameter, r_{dev}^e is the initial value of the degree of mobilization of plasticity and ε_d^p is the deviatoric plastic deformation. The parameter d represents the ratio between the preconsolidation and the critical pressure p'_{cr}, as:

$$d = \frac{p'_c}{p'_{cr}} \quad [1.18]$$

The coefficient M depends on the Lode angle ϑ to account for the effect of the stress path direction in the π plane (Figure 1.9(b)), perpendicular to the space diagonal [POT 99]. This model adopts the formulation proposed by van Eekelen [VAN 80], in which the M coefficient is defined as:

$$M = 3\sqrt{3} \cdot a_L\left(1 + b_L \cdot \sin(3\vartheta)\right)^{n_L} \quad [1.19]$$

where a_L, b_L and n_L are three material parameters. The first and second of these parameters depend on the friction angles in compression and extension, φ_c and φ_e, respectively, so that by defining:

$$r_c = \frac{1}{\sqrt{3}}\left(\frac{2 \cdot \sin\varphi_c}{3 - \sin\varphi_c}\right) \text{ and } r_e = \frac{1}{\sqrt{3}}\left(\frac{2 \cdot \sin\varphi_e}{3 + \sin\varphi_e}\right) \quad [1.20]$$

they become:

$$b_L = \frac{\left(\frac{r_c}{r_e}\right)^{1/n_L} - 1}{\left(\frac{r_c}{r_e}\right)^{1/n_L} + 1} \quad \text{and} \quad a_L = \frac{r_c}{(1 + b_L)^{n_L}}$$

$$\text{with } a_L > 0; \; b_L n_L > 0; \; -1 < b_L < 1 \qquad [1.21]$$

The value of n_L must be assumed to ensure the convexity condition [BAR 98], and the Lode angle is defined as:

$$\sin(3\vartheta) = -\left(\frac{3\sqrt{3}}{2}\frac{III_S}{II_S^3}\right) \text{ with } II_s = \frac{1}{2}s_{ij}s_{ij} \text{ and } III_s = \frac{1}{3}s_{ij}s_{jk}s_{ki} \qquad [1.22]$$

where IIs and IIIs are the second and third invariants of the deviatoric stress tensor, respectively. The flow rule is associated for the isotropic mechanism, but unassociated for the deviatoric mechanism, which means that when referring to the isotropic and deviatoric plastic potentials as g_{iso} and g_{dev} respectively, $g_{iso} = f_{iso}$ but $g_{dev} \neq f_{dev}$, as [NOV 79]:

$$g_{dev} = q - \frac{\alpha}{\alpha-1} Mp' \left[1 - \frac{1}{\alpha}\left(\frac{d \cdot p'}{p'_c}\right)^{\alpha-1}\right] = 0 \qquad [1.23]$$

where α is a material parameter that expresses the dilatancy rule:

$$\frac{d\varepsilon_v^p}{d\varepsilon_d^p} = \alpha\left(M - \frac{q}{p'}\right) \qquad [1.24]$$

The flow rules for the volumetric and deviatoric plastic strains are:

$$d\varepsilon_v^p = \lambda_{iso}^p \frac{\partial g_{iso}}{\partial p'} + \lambda_{dev}^p \frac{\partial g_{dev}}{\partial p'} \qquad [1.25]$$

$$d\varepsilon_d^p = \lambda_{dev}^p \frac{\partial g_{dev}}{\partial q} \qquad [1.26]$$

where λ_{iso}^p and λ_{dev}^p are the plastic multipliers for the isotropic and deviatoric mechanisms, respectively. These multipliers can then be calculated by starting from the consistency equation:

$$df = \frac{\partial f}{\partial \boldsymbol{\sigma}'} : d\boldsymbol{\sigma}' + \frac{\partial f}{\partial T} dT + \frac{\partial f}{\partial \boldsymbol{\varepsilon}^p} : d\boldsymbol{\varepsilon}^p \qquad [1.27]$$

where **f** is the vector that includes the two yield surfaces f_{iso} and f_{dev}. The second term on the right side of equation [1.27] is responsible for the thermoplastic component of the deformation. To account for cyclic effects, the degree of plastification for the isotropic yield mechanism changes during thermal cycles. No further thermoplastic deformation is added after a certain number of cycles, unless the maximum temperature imposed during the previous cycle is exceeded. This model reproduces the cyclic thermal accommodation phenomenon experienced by soils (Figure 1.7).

1.3.1.1. *Numerical analyses with the ACMEG-T model*

The model has been validated by Laloui *et al.* [LAL 09] and shown to be capable of reproducing the experimental results under different thermal-stress paths with good accuracy. The purpose here is to illustrate the model's performance under thermal and mechanical loading conditions similar to those imposed by energy piles. In this sense, the analysis focuses on (1) the volume variation under heating and cooling cycles and (2) the shear strength at different temperatures. The first aspect is more related to the thermally induced displacement of the foundation and the thermal effect on the stress state at the pile–soil interface, while the second aspect is more related to the pile tip response. Therefore, three examples of simulations run with the ACMEG-T model are studied to reproduce three experimental results on Boom [BAL 91] and Bangkok [ABU 06] clays. The ACMEG-T parameters of the two considered materials are listed in Table 1.2. These parameters were previously calibrated based on the experimental data available in the literature. The procedure for calibration can be found in [LAL 09]. The first case includes three tests (one NC and two OC) run on Boom clay by imposing one drained heating–cooling cycle from 20 to 95°C under constant isotropic confinement. The confinement imposed for the NC conditions was 6 MPa, while in the two OC cases, the sample was first consolidated to 6 MPa and then unloaded to 3 (OCR = 2) and 1 MPa (OCR = 6). Identical thermal-stress paths were numerically simulated with ACMEG-T, and the results, presented in Figure 1.10(a), show the ability of the model to develop thermoelastic or thermoelastoplastic deformation depending on the material's initial OCR. The second case, related to the thermal volume change of clays, concerns the test performed on Bangkok clay and already shown in Figure 1.8(a). The entire process is simulated, starting from an initial consolidation from 70 to 100 kPa, then the application of one thermal cycle (22–90–22°C) in drained conditions and the continuation of consolidation up to 200 kPa. The numerical results are shown in Figure 1.10(b) and match the experimental data, illustrating the model's ability to reproduce the thermal consolidation phenomenon discussed in this chapter.

Finally, the experimental results provided by Abuel-Naga *et al.* [ABU 06] on NC Bangkok clay were considered to check the model's ability to reproduce the responses under triaxial conditions at different temperatures. The tests are run for consolidation under an isotropic mean effective stress of 300 kPa and then sheared under an axial strain control of up to 30%. The comparison between the experimental and numerical results is shown in Figure 1.11(a) in terms of deviatoric stress and in Figure 1.11(b) in terms of volumetric deformation.

Isothermal elastic parameters			**Boom**	**Bangkok**
Bulk modulus at $p'_{ref} = 1$ MPa	K_{ref}	[MPa]	130	42
Shear modulus at $p'_{ref} = 1$ MPa	G_{ref}	[MPa]	130	15
Elastic exponent	n_e	[-]	0.4	1
Thermal elastic parameters				
Thermal expansion coefficient	β'_s	[°C^{-1}]	4×10^{-5}	2×10^{-4}
Isothermal plastic parameters				
Material parameters	a	[-]	0.007	0.02
	b	[-]	0.6	0.2
	c	[-]	0.012	0.04
	d	[-]	1.3	1.6
Friction angle in compression	φ'_c	[°]	16	22.66
Friction angle in tension	φ'_e	[°]	16	22.66
Lode parameter	n_L	[-]	–0.229	–0.229
Dilatancy parameter	α	[-]	1	2
Plastic index	β	[-]	18	5.49
Initial isotropic plastic radius	r^e_{iso}	[-]	0.001	0.15
Initial deviatoric plastic radius	r^e_{dev}	[-]	0.3	0.1
Thermal plastic parameters				
Evolution of f_{iso} with temperature	γ_T	[-]	0.2	0.22

Table 1.2. *ACMEG-T parameters for Boom and Bangkok clays*

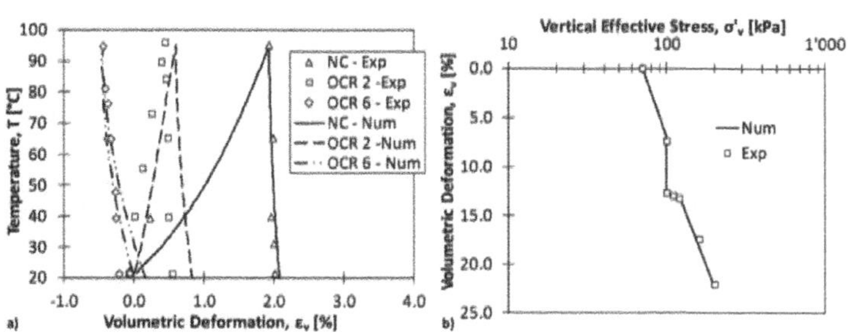

Figure 1.10. *Numerical simulations with ACMEG-T: (a) thermal deformation of Boom clay [BAL 91] and (b) thermal consolidation of Bangkok clay [ABU 06]*

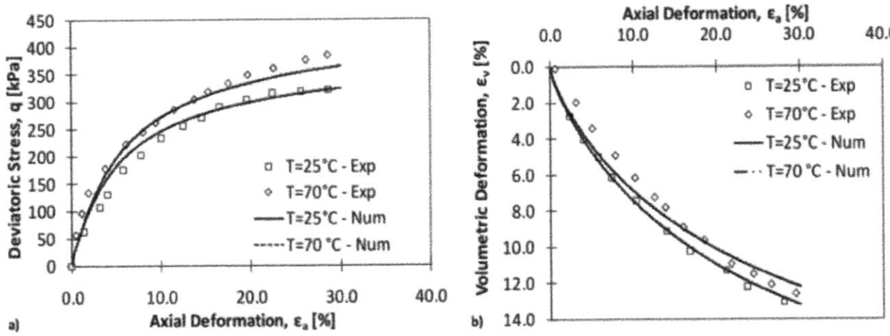

Figure 1.11. *Numerical simulations with ACMEG-T: drained triaxial tests on Bangkok clay at different temperatures [ABU 06]*

1.4. Acknowledgments

This research project was funded by the Swiss Federal Office of Energy (contract no. 154,426).

1.5. Bibliography

[ABU 06] ABUEL-NAGA H.M., BERGADO D.T., RAMANA G.V., *et al.*, "Experimental evaluation of engineering behaviour of soft Bangkok clay under elevated temperature", *Journal of Geotechnical and Geoenvironmental Engineering*, vol. 132, no. 7, 2006.

[ABU 07] ABUEL-NAGA D.T., BERGADO A., BOUAZZ A., "Thermally induced volume change and excess pore water pressure of soft Bangkok clay", *Engineering Geology*, vol. 89, pp. 144–154, 2007.

[BAL 88] BALDI G., HUECKEL T., PELLEGRINI R., "Thermal volume changes of mineral water system in low porosity clay soils", *Canadian Geotechnical Journal*, vol. 25, pp. 807–825, 1998.

[BAL 91] BALDI G., HUECKEL T., PEANO A., *et al.*, *Developments in Modelling of Thermo-Hydro-Mechanical Behaviour of Boom Clay and Clay-Based Buffer Materials (vols. 1 and 2)*, EUR 13365/1 and 13365/2, Luxembourg, 1991.

[BAR 98] BARNICHON J.D., Finite element modelling in structural and petroleum geology, Doctoral Thesis, Faculty of Applied Sciences, University of Liege, 1998.

[BOU 94] BOUDALI M., LEROUEIL S., SRINIVAS A., *et al.*, "Viscous behaviour of natural clays", *Proceedings of the 13th International Conference on Soil Mechanics and Foundation Engineering*, vol. 1, New Delhi, India, pp. 411–416, 1994.

[BUR 00] BURGHIGNOLI A., DESIDERI A., MILIZAINO S., "A laboratory study on the thermo mechanical behaviour of clayey soils", *Canadian Geotechnical Journal*, vol. 37, pp. 764–780, 2000.

[BSI 95] BSI (British Standards Institution), *Eurocode 7: Geotechnical Design*, London, 1995.

[CAM 68] CAMPANELLA R.G., MITCHELL J.K., "Influence of the temperature variations on soil behaviour", *Journal of the Soil Mechanics and Foundation Engineering Division*, vol. 94, no. SM3, pp. 709–734, 1968.

[CEK 04] CEKEREVAC C., LALOUI L., "Experimental study of thermal effects on the mechanical behaviour of a clay", *International Journal of Numerical and Analytical Method in Geomechanics*, vol. 28, no. 3, pp. 209–228, 2004.

[DAF 80] DAFALIAS Y., HERRMANN L., "A bounding surface soil plasticity model", *International Symposium on Soils Under Cyclic and Transient Loading*, Swansea, pp. 335–345, 1980.

[DEM 82] DEMARS K.R., CHARLES R.D., "Soil volume changes induced by temperature cycling", *Canadian Geotechnical Testing Journal*, vol. 19, pp. 188–194, 1982.

[DES 88] DESIDERI A.,"Determinazione sperimentale dei coefficienti di dilatazione termica delle argille", *Proceedings Convengo del Gruppo Nazionale di Coordinamento per gli Studi di Ingegneria Geotecnica*, vol. 1, *sul tema: Deformazioni dei terreni ed interazione terreno-struttura in condizioni di esercizio*, Monselice, Italy, pp. 193–206, 5–6 October 1988.

[DI 13] DI DONNA A., DUPRAY F., LALOUI L. "Numerical study of the effects induced by thermal cyclic soil plasticity on the geotechnical design of energy piles", *Computers and Geotechnics*, 2013, in press.

[DIX 93] DIXON D., GRAY M., LINGNANU B., *et al.*, "Thermal expansion testing to determine the influence of pore water structure on water flow through dense clays", *46th Canadian Geotechnical Conference*, Saskatoon, pp. 177–184, 1993.

[ERI 89] ERIKSSON L.G., "Temperature effects on consolidation properties of sulphide clays", *12th International Conference on Soil Mechanics and Foundation Engineering*, vol. 3, Rio de Janeiro, pp. 2087–2090, 1989.

[FRA 08] FRANCOIS B., LALOUI L., "ACMEG-TS: a constitutive model for unsaturated soils under non-isothermal conditions", *International Journal of Numerical and Analytical Methods in Geomechanics*, vol. 32, pp. 1955–1988, 2008.

[HON 13] HONG P.Y., PEREIRA J.M., TANG A.M., *et al.*, "On some advanced thermo-mechanical models for saturated clays", *International Journal for Numerical and Analytical Methods in Geomechanics*, 2013. DOI: 10.1002/nag.2170.

[HUE 90] HUECKEL T., BALDI G., "Thermo plasticity of saturated clays: experimental constitutive study", *Journal of Geotechnical Engineering*, vol. 116, pp. 1778–1796, 1990.

[HUE 92] HUECKEL T., "On effective stress concepts and deformation in clays subjected to environmental loads: discussion", *Canadian Geotechnical Journal*, vol. 29, pp. 1120–1125, 1992.

[HUE 98] HUECKEL T., PELLEGRINI R., DEL OLMO C., "A constitutive study of thermo-elasto-plasticity of deep carbonatic clays", *International Journal of Numerical and Analytical Methods in Geomechanics*, vol. 22, pp. 549–574, 1998.

[HUE 09] HUECKEL T., FRANCOIS B., LALOUI L., "Explaining thermal failure in saturated clays", *Géotechnique*, vol. 59, no. 3, pp. 197–212, 2009.

[HUJ 79] HUJEUX J.C., Calcul numérique de problèmes de consolidation élastoplastique, Doctoral Thesis, Ecole Centrale, Paris, 1979.

[LAL 93] LALOUI L., Modélisation du comportement thermo-hydro-mécanique des milieux poreux anélastique, Doctoral Thesis, Ecole Centrale de Paris, 1993.

[LAL 03] LALOUI L., CEKEREVAC C., "Thermo-plasticity of clays: an isotropic yield mechanism", *Computers and Geotechnics*, vol. 30, no. 8, pp. 649–660, 2003.

[LAL 08] LALOUI L., CEKEREVAC C., "Non-isothermal plasticity model for cyclic behaviour of soils", *International Journal for Numerical and Analytical Methods in Geomechanics*, vol. 32, no. 5, pp. 437–460, 2008.

[LAL 09] LALOUI L., FRANÇOIS B., "ACMEG-T: soil thermoplasticity model", *Journal of Engineering Mechanics*, vol. 135, no. 9, pp. 932–944, 2009.

[LAN 99] LANCELLOTTA R., CALAVERA J., *Fondazioni*, McGraw-Hill, 1999.

[MCK 65] MCKINSTRY H.A., "Thermal expansion of clay materials", *American Mineralogist*, vol. 50, pp. 210–222, 1965.

[MIL 92] MILIZIANO S., Influenza della temperatura sul comportamento meccanico delle terre coesive, Doctoral Thesis, University of Rome "La Sapienza", Italy, 1992.

[MOD 97] MODARESSI H., LALOUI L., "A thermo-viscoplastic constitutive model for clays", *International Journal for Numerical and Analytical Methods in Geomechanics*, vol. 21, no. 5, pp. 313–315, 1997.

[MOR 95] MORITZ L., Geotechnical properties of clay at elevated temperatures, Report: 47, Swedish Geotechnical Institute, Linköping, 1995.

[NOV 79] NOVA R., WOOD D.M., "Constitutive model for sand in triaxial compression", *International Journal for Numerical and Analytical Methods in Geomechanics*, vol. 3, no. 3, pp. 255–278, 1979.

[PLU 69] PLUM R.L., ESRIG M.I., Some temperature effects on soil compressibility and pore water pressure. Effect of temperature and heat on engineering behaviour of soils, Highway research board special report, vol. 103, pp. 231–242, 1969.

[POT 99] POTTS D.M., ZDRAVKOVIĆ L., *Finite Element Analysis in Geotechnical Engineering: Theory*, Thomas Telford Limited, 1999.

[TID 89] TIDFORS M., SÄLLFORS S., "Temperature effect on preconsolidation pressure", *Geotechnical Testing Journal*, vol. 12, no. 1, pp. 93–97, 1989.

[TOW 93] TOWHATA I., KUNTIWATTANAUL P., SEKO I., *et al.*, "Volume change of clays induced by heating as observed in consolidation tests", *Soil and Foundations*, vol. 33, pp. 170–183, 1993.

[VAN 80] VAN EEKELEN H.A.M., "Isotropic yield surfaces in three dimensions for use in soil mechanics", *International Journal for Numerical and Analytical Methods in Geomechanics*, vol. 4, no. 1, pp. 89–101, 1980.

Chapter 2

Full-scale *In Situ* Testing of Energy Piles

This chapter deals with *in situ* investigations of the hermomechanical response of full-scale energy piles. First, conventional sensors utilized to monitor strains, stresses and temperature in the piles and in the soil are described as well as the data treatment procedures. Then, two major *in* situ experimental sites built on the Ecole Polytechnique Fédérale de Lausanne (EPFL) campus are described. The contributions of these sites to the understanding of the thermomechanical behavior of energy piles are discussed. The first site was dedicated to the thermomechanical response of one single pile while the second site was built to investigate the interactions occurring within a group of energy piles.

2.1. Monitoring the thermomechanical response of energy piles

2.1.1. *Measuring strains and temperature along the piles*

Measuring strains in piles for our *in situ* experiments is achieved with two types of sensor: optical fibers and strain gauges. Optical fibers provide accurate strain measurements but require more logistics (i.e. a large dedicated reading unit plus a computer to operate it) while strain gauges provide reliable measurements of strain and temperature with a small, simple reading unit.

The systems used on the EPFL campus to measure strains in the piles are the following:

– The *Surveillance d'Ouvrages par Fibres Optiques* (SOFO) system of optical fibers from SMARTEC^{TM}/ROCTEST^{TM}. This system is based on the comparison of

---

Chapter written by Thomas MIMOUNI and Lyesse LALOUI.

a signal traveling along a reference fiber and a measuring fiber. The reference fiber is long enough so that it is never put under tension. As a result, strains on the reference fiber are only caused by temperature variations: the system is auto-corrected for temperature changes [GLI 00, LLO 00, INA 00]. The system directly provides information on the deformations and therefore does not need further data treatment.

– Vibrating wire strain gauges, model C 110 from TELEMAC™ located at the first site and model EM-5 from ROCTEST™ situated on the second site are used to measure the strain. As the tests are carried out in non-isothermal conditions, some of the strain of the vibrating wire is due to temperature changes. Therefore, a 3 kΩ thermistor is included in each strain gauge to correct temperature change. The gauge outputs are the resonant frequency of the vibrating wire and its temperature. Therefore, data processing is required.

These two types of sensor were deployed along the pile axes, attached to the reinforcing cages. Since optical fibers are delivered with anchors already mounted on the sensor body, tie wires or plastic ties were used to attach them along the rebars of the reinforcing cages (Figure 2.1). However, fibers must be attached with an initial tension to avoid having loose fibers after the mechanical compression of piles. Strain gauges were attached to frames prior to their installation on the reinforcing cages with a wire attached perpendicular to the gauge axes, which avoids measuring the axial strain of the frame. The frames were designed to fit exactly in between the rebars of the reinforcing cages. Next, the frames were attached to the reinforcing cages with wire ties (Figure 2.2); this saved a substantial amount of time onsite and provided a reliable attachment of the sensors.

Data processing for the strain gauges is based on the vibrating wire theory. The actual resonant frequency of a wire is compared to a reference value and this provides information about its strain ε. Indeed, the resonant frequency F of a wire under tension is given by:

$$F = \frac{1}{2L}\sqrt{\frac{T}{\mu}} \qquad [2.1]$$

where L is the string length (in m), T is the string tension (in N) and μ is the string linear density (in kg/m). The wire tension can be related to its strain ε through its Young's modulus E_w and cross-section A by:

$$T = AE_w\varepsilon \qquad [2.2]$$

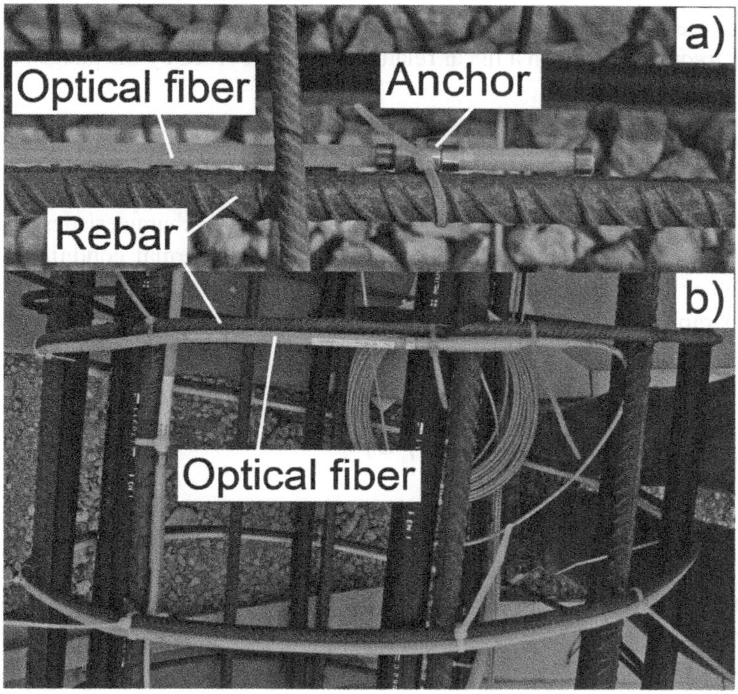

Figure 2.1. *Photographs of optical fibers attached to the reinforcing cages of the piles (on the second test site). One fiber a) is attached along the pile axis and the other b) is attached along a reinforcing hoop to monitor radial strains*

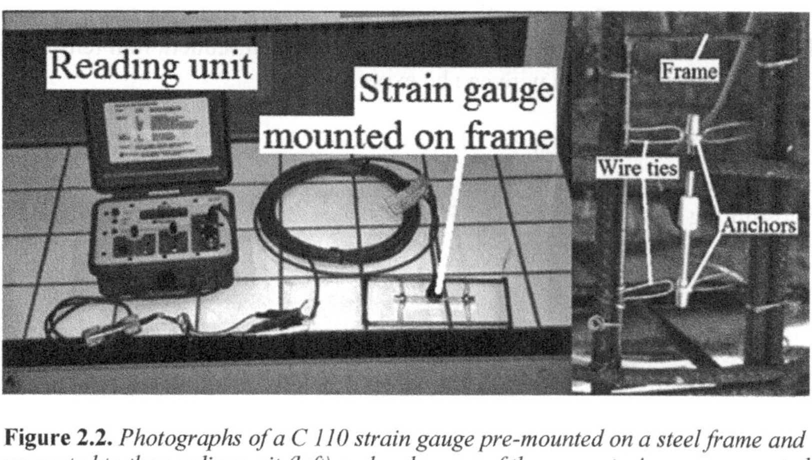

Figure 2.2. *Photographs of a C 110 strain gauge pre-mounted on a steel frame and connected to the reading unit (left) and a close-up of the same strain gauge mounted on the reinforcing cage (right)*

Therefore, the strain in the wire can be directly linked to the square of its resonant frequency through a linear relationship as provided by the gauge supplier:

$$\varepsilon = \frac{K}{1{,}000} F^2 \qquad [2.3]$$

where K is a constant depending on the instrument characteristics. However, because the experiments are carried out under non-isothermal conditions, thermal expansion or contraction of the wire must be accounted for. The thermal strain of the wire is simply superimposed on to the measured strain, knowing that the linear thermal expansion coefficient of the wire α_w^T is 11.5 $\mu\varepsilon$/°C:

$$\Delta\varepsilon_{\mathbf{obs}} = \frac{K}{1{,}000}(F_1^2 - F_0^2) + \alpha_w^T (T_1 - T_0) \qquad [2.4]$$

where F_1 and F_0 are the actual and reference frequencies, respectively, and T_1 and T_0 are the actual and reference temperatures, respectively. Any increase in the temperature of the wire induces expansion and consequently reduces its tension. Therefore, the observed tension is lower than the mechanically induced tension and can be written as:

$$\Delta T_{\mathbf{obs}} = \Delta T_{\mathbf{mech}} + \Delta T_{\mathbf{th}} \qquad [2.5]$$

where ΔT_{obs} is the measured variation in tension linked to the measured variation in frequency. ΔT_{mech} and ΔT_{th} are the mechanical and thermal variations of tension in the wire, respectively. ΔT_{obs} is estimated by multiplying equation [2.3] by Young's modulus and cross-section of the wire, while the thermal reduction in tension due to the thermal expansion of the wire can be expressed by:

$$\Delta T_{\mathbf{th}} = -AE\alpha_w^T (T_1 - T_0) \qquad [2.6]$$

Substituting this into equation [2.5] yields:

$$\Delta T_{mech} = AE\frac{K}{1{,}000}\left(F_1^2 - F_0^2\right) + AE\alpha_w^T\left(T_1 - T_0\right) \qquad [2.7]$$

Dividing equation [2.7] by Young's modulus and the cross-section of the wire yields equation [2.4].

The vibrating wire reading unit, model MB-3TL from ROCTEST™, provides measurements of the vibration frequency to within an accuracy of ±0.5%

FS (Full-Scale). However, the accuracy of the optical fibers is 0.2% FS. The uncertainty in strain gauge reading is ±0.05 Hz for the frequency and ± 0.05°C for the temperature. As a result, the uncertainty in the strain measurement, $\delta\varepsilon$, is given by:

$$\delta\varepsilon = \frac{K}{1{,}000}(2F\delta F + \delta F^2) + \alpha_w^T \delta T \qquad [2.8]$$

The resolution of the measurements for an EM-5 stain gauge (K~4) for a maximum frequency of 1,200 Hz is, therefore, equal to 1 µε.

2.1.2. *Measuring pile tip compression*

Pile tip compression is monitored using model HVC pressure cells from SMARTEC™ on the first site and Total Pressure Cell (TPC) pressure cells from ROCTEST™ on the second site. The cells installed on the *in situ* sites at EPFL are coupled with vibrating wire transducers, whose principle is the same as for strain gauges and a 3 kΩ thermistor is used for temperature correction.

The cells were attached to a cross that was welded at the base of the reinforcing cages (Figure 2.3). However, the installation of this type of cell is challenging when intended for pile tip monitoring. Indeed, shrinkage of the concrete might degrade the contact between the cell and the surrounding concrete. Pressurization tubes are available to inflate the cell when concrete shrinkage is deemed to be significant but having such a tube along the whole pile axis can present several additional challenges. Furthermore, precasting the cells within a mass of concrete and then inflating them before installing the block at the tip of the reinforcing cages would have been time consuming. The solution adopted to overcome these difficulties was pouring some concrete into the bottom of the boreholes using a pit tube and pushing the reinforcing cages, with the cells attached at their base, into the fresh concrete. Although this method was successful on piles #1, #3 and #4, the cell at the tip of pile #2 (Figure 2.7) did not record any pressure changes during the tests. This was attributed to the fact that this reinforcing cage was first dropped and then lifted and maintained in this position because the borehole was deeper than expected.

2.1.3. *Monitoring the behavior of the soil*

At the second test site, the thermomechanical response of a group of energy piles was investigated using sensors to monitor the pore water pressure, soil displacements and temperature in the soil between the piles.

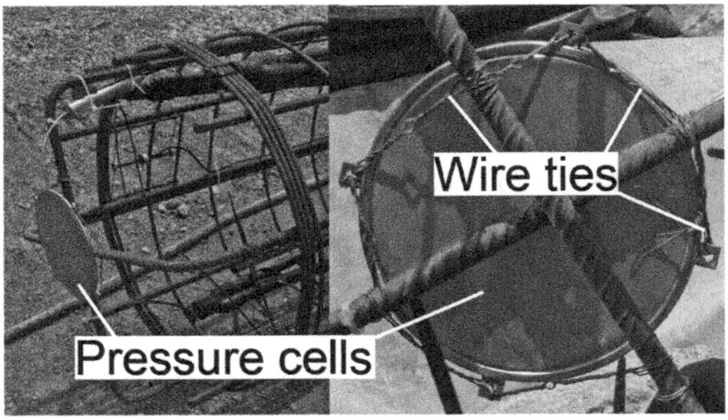

Figure 2.3. *Pressure cells attached at the base of reinforcing cages on the second test site*

Soil displacements were monitored using two BOR-EX borehole extensometers from ROCTESTTM. These were designed to be used with three rebar anchors at different depths (−15 m, −12 m and –5 m). The three anchors are linked to a reference plate by stainless steel rods. The borehole extensometers were installed in boreholes with a diameter of 160 mm and then backfilled with an 80/20 cement-bentonite mixture [MIK 02]. Reference plates were grouted into the concrete of the piled raft to act as the reference. Measurements in the extensometer head (Figure 2.4) are achieved using vibrating wire transducers.

Piezometers (Figure 2.4) were installed within two dedicated boreholes of 160 mm diameter. Once the desired borehole depth was reached, a PVC screen pipe was inserted into the temporary casing. Then, the sensors were installed within the screen pipe: piezometers were installed within layers of sand sealed with compactonite, a highly expansive clay material. One piezometer was installed at the bottom of each borehole, about 17 m below the soil surface, and another at mid-height. The piezometers are coupled with vibrating wire transducers so they are equipped with 3 kΩ thermistors for temperature correction. Additional thermistors (TH-T model from ROCTEST^{TM}) were also installed in-between the piezometers to provide full temperature profiles in the ground.

2.2. Description of the two full-scale *in situ* experimental sites

2.2.1. *Single full-scale test pile*

A pile supporting a four-storey building (the BP building at EPFL) was chosen as a test pile. It is located below one edge of the 100 m long and 30 m wide building. The pile is 0.88 m in diameter and 25.8 m in length and is embedded within a layered soil that consists of two alluvial layers (from 0 to –5.5 m and from –5.5 to –12 m) and two moraine layers (from –12 to –22 m and from –22 to –25 m) and sits in the sandstone bedrock, which extends from –22 m and below the pile tip (Figure 2.5). As a result, the test pile is of the semi-floating type because it transfers its load to the ground through friction (in the moraine and alluvial layers) and base compression (in the sandstone bedrock).

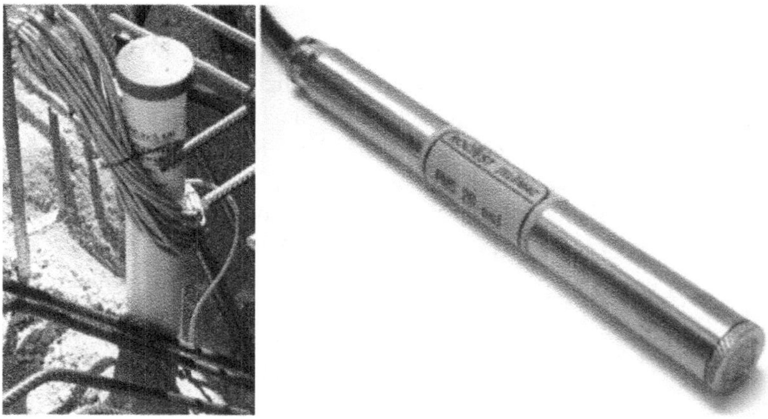

a) b)

Figure 2.4. *a) Reading head of a borehole extensometer embedded within the piled raft and b) the PWS piezometer from ROCTEST™ installed on the second test site*

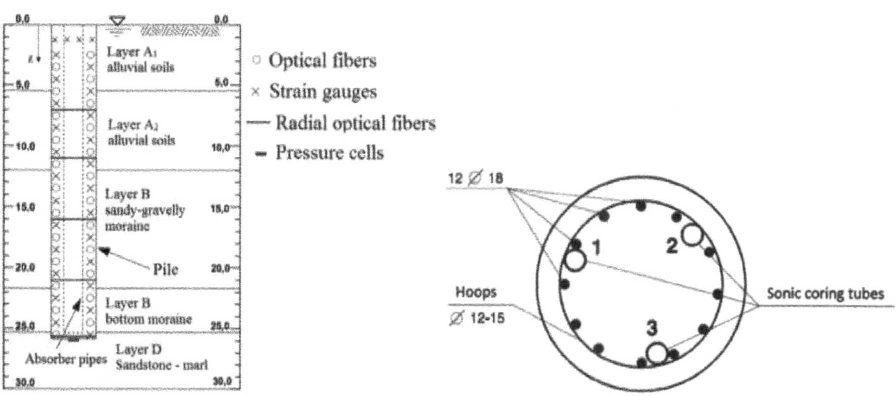

a) b)

Figure 2.5. *a) Stratigraphy and b) cross-section of the first test, from [LAL 03]*

The test pile is a bored pile, where a temporary casing was first pushed into the ground until it rested on top of the sandstone bedrock. Then, the space within the temporary casing was excavated with a helical auger. Pile socketing was achieved with a cylindrical bucket auger. Next, the reinforcing cage was inserted into the temporary casing and concrete was poured using a pit tube. The temporary casing was finally removed by pulling and further concrete was poured to adjust the concrete level in the borehole.

The reinforcing cage of the pile consists of 12 rebars, 18 mm in diameter, tightened together with 15 mm diameter reinforcing hoops spaced 15 cm apart. Three vertical tubes were also installed inside the reinforcing cage for the pile integrity test (PIT) and sonic coring (Figure 2.5). Results of the PIT show an enlargement of the pile diameter just above –20 m (Figure 2.6). Sonic coring was used to estimate Young's modulus of the concrete E_C based on the sound velocity C within the pile given by [STR 91]:

$$C = \sqrt{\frac{E_c(1-\nu)}{\rho_c(1+\nu)(1-2\nu)}} \qquad [2.9]$$

where ρ_c and ν are the density and Poisson's coefficient of the concrete ($\rho_c \sim$ 2,500 kg/m^3), respectively. The three tubes allowed three different profiles of the sound velocity (tube 1→2, tube 1→3 and tube 2→3) to be obtained and, therefore, of the concrete rigidity. Estimates of E_C were achieved assuming values for Poisson's coefficient of 0 and 0.16. Measurements were taken at different dates with different temperatures (Table 2.1). The linear elastic modulus of the pile, E_P, is derived from these estimates using:

$$E_P = E_C\left[1+\varphi\frac{E_R}{E_C}\right] \qquad [2.10]$$

where E_R is the elastic modulus of the steel of the reinforcing cage and φ is the percentage of reinforcing rods. A value of $E_P = 29{,}200$ MPa (with $\nu = 0$) was finally selected for the rest of the study.

The pile was equipped with absorber pipes shaped in four U-loops connected in parallel to collectors at the pile head. The loops are made of high-density polyethylene (HDPE), which is also used for geothermal loops. The pile has no heating or cooling purpose but, during the tests, the absorber pipes are connected to a mini-module that is equipped with heaters and circulation pumps to simulate the effects of thermal loading.

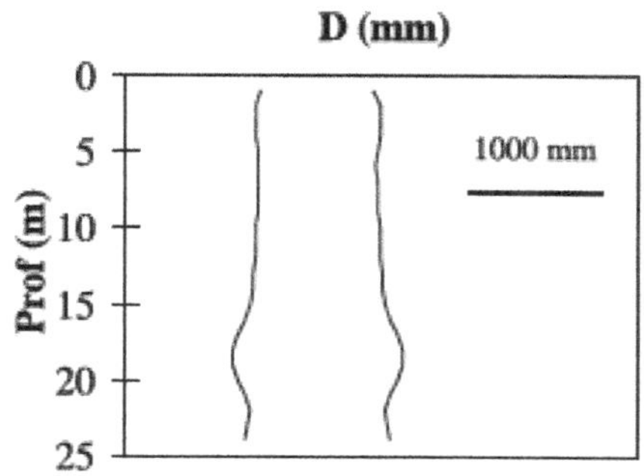

Figure 2.6. *Results of the PIT carried out on the test pile, from [LAL 99]*

Date (mm-dd-yyyy)	Mean pile temperature (°C)	E_C (MPa) ($\nu = 0$)	E_C (MPa) ($\nu = 0.16$)
02-24-1998	23	28,068	26,357
05-25-1998	35	27,206	25,547
06-03-1998	24	31,097	29,201
05-25-1999	19	31,948	30,000

Table 2.1. *Sonic coring results, from [LAL 03]*

2.2.2. *Full-scale test on a group of energy piles*

Four full-scale test piles were installed below one corner of a raft supporting a 9 m × 25 m water retention tank on top of a heavily reinforced 0.9 m thick raft. This raft is also founded on conventional piles and the test piles are located within a 4.21 m × 4.21 m square below the north-western corner of the tank. No conventional pile is located between the test piles, which are arranged in a triangle so that the distance between pile #1 and the other piles is 3 m, while the distance between pile #2 and piles #3 and #4 is 4.21 m (Figure 2.7).

The compressive strength and critical stress of the concrete were estimated from compression tests on cylindrical samples (16 mm in diameter and 32 mm high) that were taken from piles #1, #2 and #3 (two samples per pile). The results of the compressive tests are listed in Table 2.2.

The piles were bored and cast onsite. A temporary casing was installed prior to boring to maintain the borehole shaft. This casing crossed the moraine and alluvial layers and rested on the sandstone bedrock. Excavation of the soil was achieved using an auger through the alluvial and moraine layers and a drilling bucket through the sandstone bedrock. The reinforcing cages for the piles were made up of 10 vertical rebars with a diameter of 18 mm and maintained by 9 mm diameter reinforcing hoops spaced at 0.2 m intervals. These reinforcing cages, with a diameter of 0.76 m, were installed into boreholes that are 0.9 m in diameter and 28 m deep.

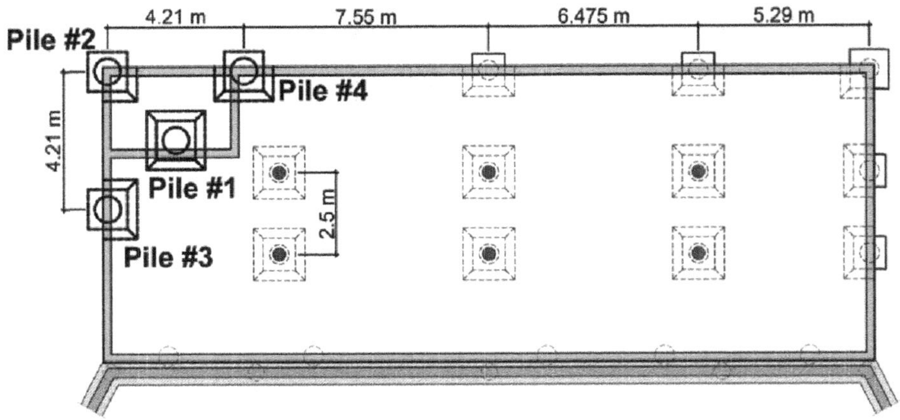

Figure 2.7. *Schematic of the water retention tank and the experimental group of piles*

Sample	1a	1b	2a	2b	3a	3b
E (MPa)	28,000	27,100	26,000	21,100	23,300	32,400
σ_c (MPa)	41.0	40.5	44.7	30.3	57.9	55.2
ρ (kg/m^3)	2,440	2,460	2,450	2,460	2,450	2,450

Table 2.2. *Results of compression tests on concrete samples. Two samples (a and b) were taken from each of the piles #1, #2 and #3*

Absorber pipes were attached on the inner face of the reinforcing cages with large plastic ties prior to installation. These consisted of four geothermal loops (per pile) connected in series and deployed all around the inner surface of the cages. The top of the U-loops is placed 4 m below the pile head and the inlet and outlet of the circuit are thermally insulated in order to prevent thermal interactions with the water retention tank above (Figure 2.8).

2.2.3. *Testing procedure*

2.2.3.1. *Mechanical loading*

The tests carried out on the single test pile located below the BP building of EPFL (first test site) followed the construction sequence in order to monitor the influence of the number of floors built on the pile confinement. As a result, seven separate tests were carried out, corresponding to the construction of the seven floors (Figure 2.9).

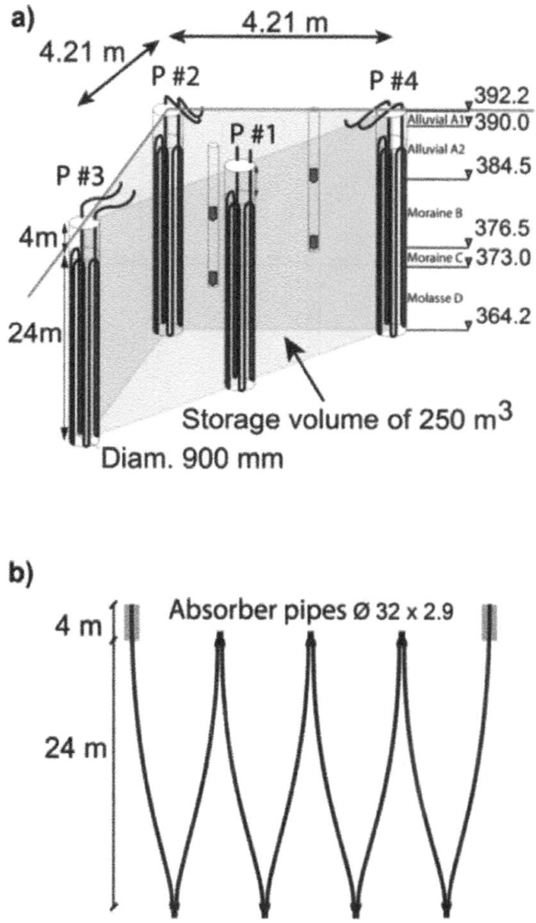

Figure 2.8. *a) Stratigraphy of the location where the pile group is embedded and b) absorber pipes. The first 4 m of the absorber pipes, just below the pile head, was thermally insulated to prevent thermal interactions with the water retention tank*

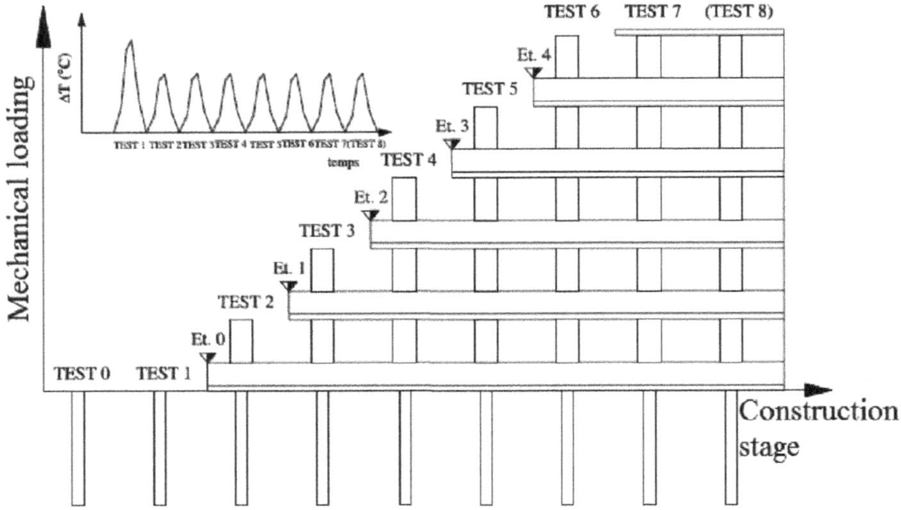

Figure 2.9. *Schematic of the different tests carried out during the construction of the building, from [LAL 03]*

Mechanical loads on the piles at the second test site did not vary since only the water retention tank was supported. The loads carried by the piles are as follows: pile #1 carries no load; pile #2 carries about 800 kN load while piles #3 and #4 carry 2100 kN load. Even though the mechanical loads applied to the piles are constant, two different types of test were carried out on this group: the first type of test (given in detail later in this chapter) consisted of heating one of the four piles at a time and estimating its profile of degree of freedom (similar to the tests carried out on the first site). The second test consisted of heating the other three piles in the group prior to heating the pile of interest. This second type of test may provide insight into the interactions between the piles that occur at the pile heads through the raft and supported structure. This kind of interaction can have a significant impact on the differential settlements and internal stresses observed within foundations where only a part of the piles is turned into heat exchangers [DUP 13].

2.2.3.2. *Thermal loading*

The piles are heated using a mini-module designed for thermal response tests (TRTs). This device fits in a flightcase and allows the delivery of a heating power from 0 to 9 kW that is adjustable in steps of 1 kW. The mini-module is equipped with two heaters and one expansive vessel. A pump is dedicated to the filling of the absorber pipes and pressurization of the circuit. Another pump is used to circulate the heat carrier fluid [MAT 08]. The module requires a 380 V power source with 16 or 32 A, depending on the power level reached during the tests. An electrical transformer installed within the module provides 220 V/10 A of electricity to the sensors and data logger. Electrical outlets are also powered by the electrical transformer, allowing the user to connect a computer or lights (Figure 2.10).

a)

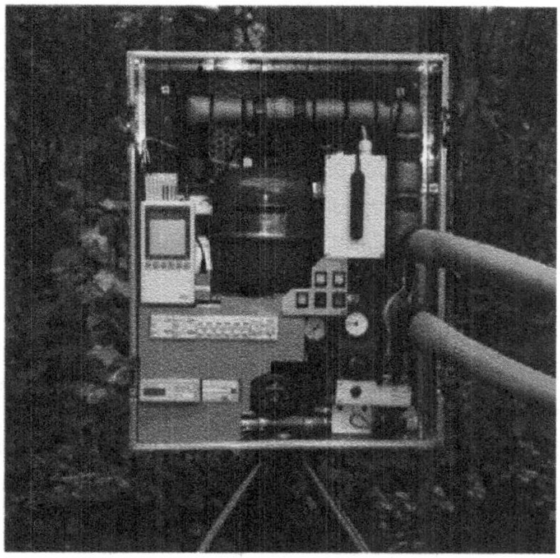

b)

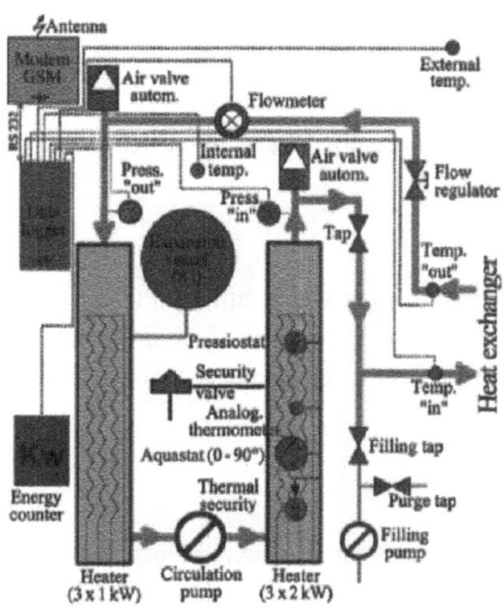

Figure 2.10. *a) Photograph of the mini-module) and b) schematic of the circuits and sensors, from [MAT 08]*

The procedure adopted on the piles is the same as that used for TRTs on geothermal loops. The mini-module is first connected to the absorber pipes embedded in the pile. Next, the circuit formed by the absorber pipes and the mini-module is filled and pressurized to 3 bars. The circuit is then purged to remove the air trapped when the mini-module is connected to the pile. Finally, the flow rate of the heat-carrier fluid is adjusted to about 25 L/min and the heaters are activated. Pile cooling is passive and is achieved by allowing the pile to rest by turning off the heaters but conserving the flow within the absorber pipes. Other reported tests have utilized heat pumps connected to a heat sink pile in order to achieve active cooling [BOU 09].

Pressuremeters and thermocouples were installed at the inlet and outlet of the module and all the sensors were connected to a data logger. The heat flow P_{abs} (in W) absorbed by the pile can be estimated with:

$$P_{abs} = Q\rho_w c_w \left(T_{in} - T_{out}\right) \quad [2.11]$$

where Q is flow rate of the heat-carrier fluid (in m^3/s), ρ_w and c_w are the volumetric mass (in kg/m^3) and specific heat (in J/kg/K) of the heat-carrier fluid and $T_{in} - T_{out}$ is the temperature difference between the inlet and outlet of the pile (in K). An energy counter was used to monitor the electricity consumption of the module during the tests.

Temperature variations imposed on the single test pile reached extreme values up to +20°C while increases of only +3°C to +5°C were initially tested on the group of piles.

2.3. Thermomechanical behavior of energy piles

2.3.1. *General methodology*

The behavior of the test piles was quantified in terms of the degree of freedom n, comparing the ability of the pile to expand or contract under thermal loading to its free thermal strain. The degree of freedom is given by:

$$n = \frac{\Delta\varepsilon_{obs}}{\Delta\varepsilon_{free}} \quad [2.12]$$

where $\Delta\varepsilon_{obs}$ is the measured strain given by equation [2.7] and $\Delta\varepsilon_{free}$ is the free strain given by:

$$\Delta\varepsilon_{free} = \alpha_C^T \Delta T \quad [2.13]$$

where α_C^T is the thermal expansion coefficient of the concrete, equal to 10^{-5}°C^{-1}. Part of the free strain is effectively observed while the remaining part is converted into the internal thermal stress σ_{th}, given by:

$$\sigma_{th} = -E_C\left(1-n\right)\alpha_C^T \Delta T \qquad [2.14]$$

where the minus sign represents the convention of negative compression.

In practice, the product $n\alpha_C^T$ is estimated for each strain gauge (i.e. at different depths) using linear regressions on curves of observed strain versus temperature (Figure 2.11).

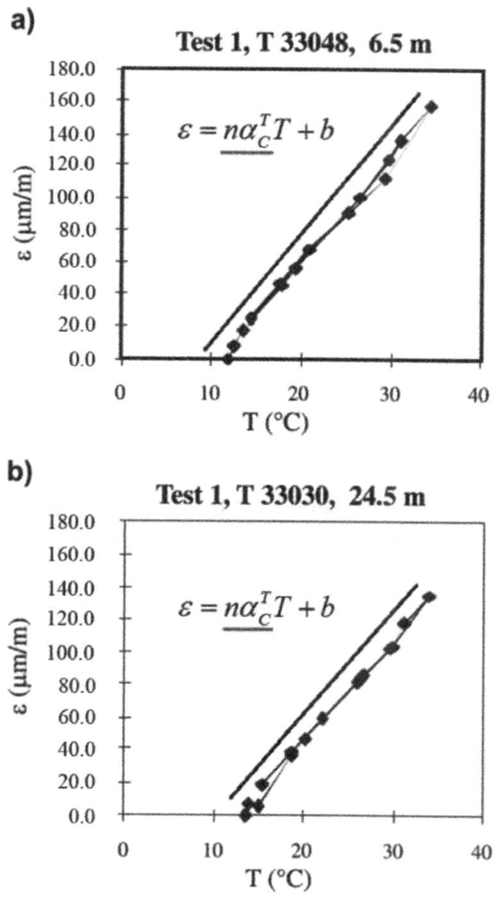

Figure 2.11. *Determination of the degree of freedom n from measurements obtained from strain gauges located at a) 6.5 m and b) 24.5 m depths, from [LAL 03]*

2.3.2. *Thermomechanical response of the single test pile*

Evolution of temperature in the single pile during test number 7 is shown in Figure 2.12 as an example. The temperature profile of the pile during heating is homogeneous. Conversely, natural cooling deforms this profile, showing higher temperatures in the center of the pile.

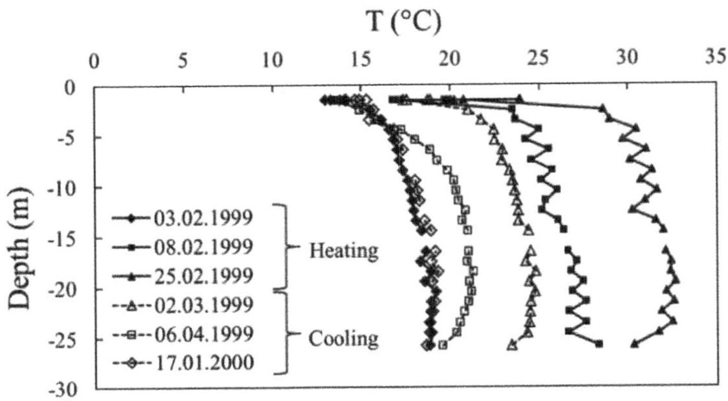

Figure 2.12. *Evolution of the temperature profile along the pile during test 7 on the single test pile at EPFL, redrawn from [LAL 09]*

At the end of the cooling period, the initial and final profiles match. As a result, the contraction of the pile is gradual and starts from the pile ends.

Evolution of the constraints acting on the test pile indicates that the greatest modification occurred during the construction of the first floor. The degree of freedom of the pile reached about 0.5 when building construction was completed. The base of the pile seems less restrained than the rest of the foundation (Figure 2.13). This phenomenon is reproduced by the design tool Thermo-Pile [KNE 11] and with finite element analyses [DUP 13], and has yet to be clearly explained.

Measurements from the stain gauges suggest that the pile strains remain thermoelastic [LAL 03].

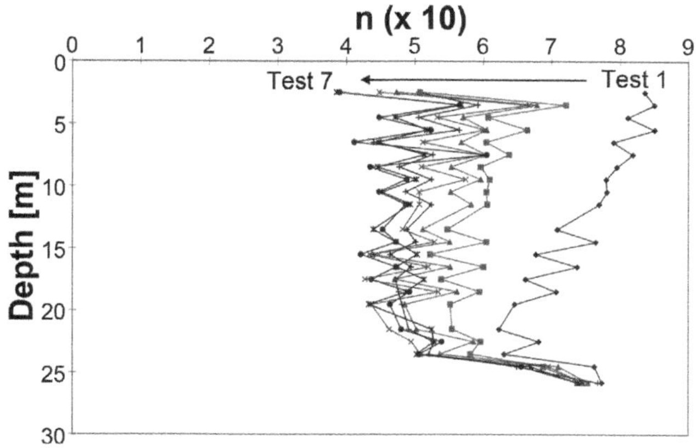

Figure 2.13. *Evolution of the profile of the degree of freedom of the single test pile during the building construction, from [LAL 99]*

Thermally induced internal stresses that are estimated using equation [2.14] can be superimposed on to the mechanical stress in order to obtain the profile of the total load transmitted along the pile axis. The maximum increase in internal load during test 7 and for a temperature increase of 13.4°C is plotted in Figure 2.14. This increase reaches 1,000 kN at the pile head while it is around 2,000 kN at the pile tip. On average, an increase in temperature of 1°C leads to an approximate increase in the pile internal load of 100 kN (i.e. 165 kPa/°C).

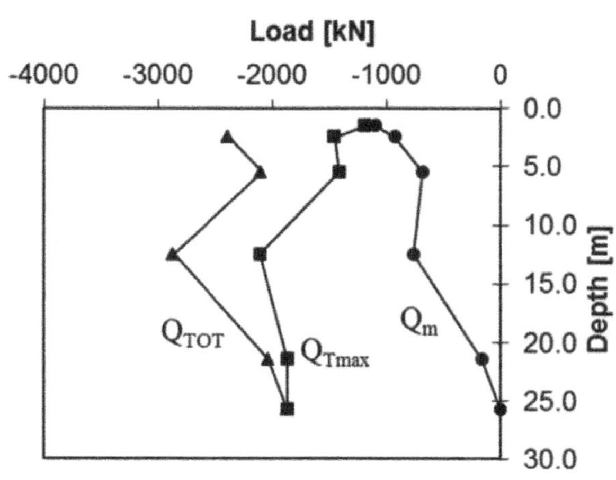

Figure 2.14. *Profiles of mechanical (Qm) and thermal (Q_T) loads during test 7 for a mean temperature increase in the pile of 13.4°C, from [LAL 03]*

2.3.3. *Thermomechanical response of a group of energy piles*

An example of the evolution of temperature during the tests is given along pile #1 in Figure 2.15. The temperature remains homogeneous along the axis of the pile except at the pile top (insulated part) and pile base.

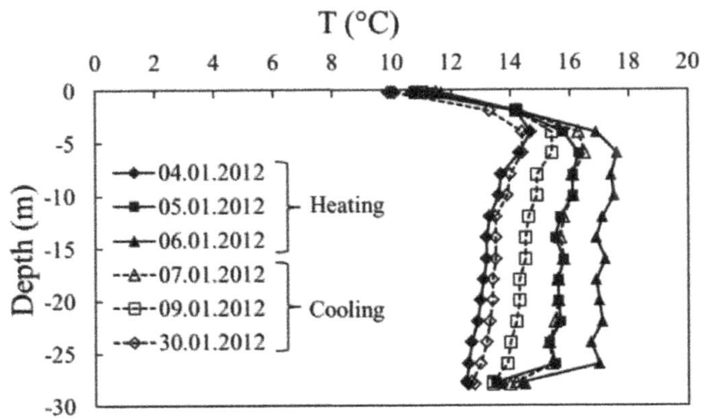

Figure 2.15. *Evolution of the temperature profile along the pile #1 during the test after the tank construction*

The goal of these first tests was to quantify the impact of the tank construction on the pile confinements. To do so, the profile of the degree of freedom of pile #1 was first determined without any structure above it ("free head" in Figure 2.16) and this was taken as the reference. Then, each pile was tested after the construction of the tank.

Comparisons between the reference profile ("free head") and the profiles measured after the tank construction suggest that pile #1 is the most affected by the construction loading (Figure 2.16). However, it is also the pile that carries the lowest load (section 2.2.3). This shows the influence that the pile location below the raft has on its confinement. Indeed, pile #1 is below an important section of raft and it is surrounded by other piles. As a result, the capping force developed by the raft as a reaction to pile expansion is the most important on pile #1. Similarly, piles #3 and #4 are located below the edges of the raft and therefore undergo greater reactions from the raft during expansion than pile #2, which is below a corner of the raft (Figure 2.7).

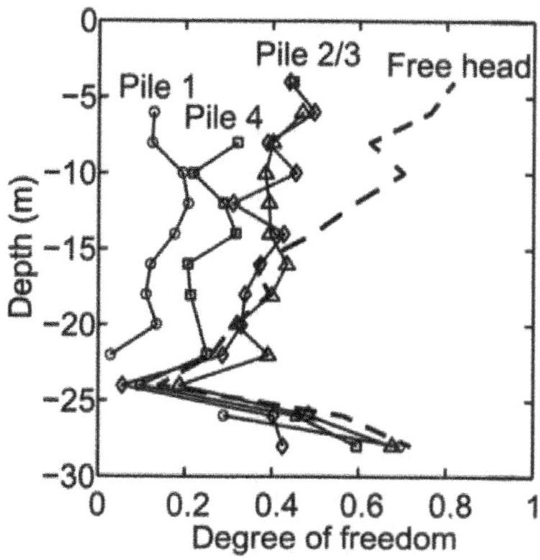

Figure 2.16. *Evolution of the profiles of the degree of freedom before and after tank construction*

Furthermore, because pile #4 is surrounded by two static piles, it is more constrained than pile #3 even though these two piles carry similar loads. As a result of these constraints, an increase of 1°C in the pile temperature leads to a maximum increase in internal load of 40–50 kN.

Measurements from the pressure cell at the tip of pile #1 indicate that the base compression remained elastic during the second test (with the tank on top of the piles) with temperature variations in the pile ranging from +2°C to +4°C (Figure 2.17).

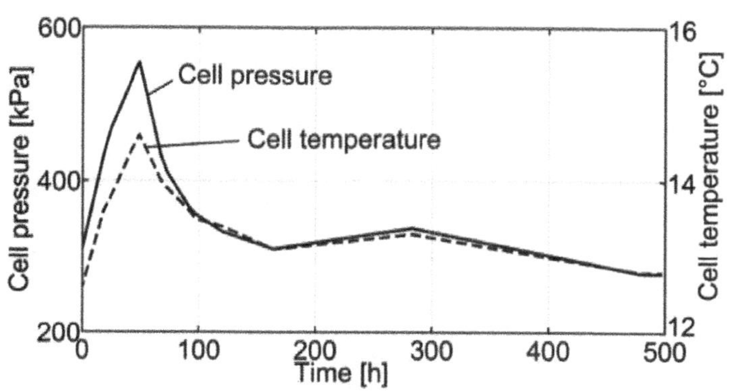

Figure 2.17. *Evolution of pressure at the tip of pile #1 after the construction of the water retention tank*

2.4. Conclusions

Full-scale *in situ* testing remains a major tool for investigating the thermomechanical response of energy piles because they can undergo various loads in many different conditions. However, full-scale experiments are expensive in terms of cost and time. Field campaigns often require at least 1 month of effective work to collect data and the same amount of time to process and correct the data. Extensive and detailed notes should be made during field campaigns since unexpected events are likely to occur (e.g. electrical supply problems, heat-carrier fluid leaks, etc.); such steps will help in removing erratic and/or erroneous data.

Sensors dedicated to this type of *in situ* application need to be robust; however, particular attention is required when handling the reinforcing cages, from their fabrication to their insertion into the boreholes. Indeed, lifting the cages can bend them and break the ties that keep the sensors or absorber pipes in position. Concrete pouring should always be achieved with a pit tube in order to prevent any sensor tearing.

Data obtained from full-scale *in situ* tests have shown that the behavior of the pile remains thermoelastic. The influence of the structure was highlighted by the two experimental sites built on the EPFL campus. Observations made on the first test pile have shown that the impact of the structure on the pile confinement does have an effect, which was observed along almost the entire pile. The main part of the constraint was observed once the raft and the first floor were built. Measurements at the second test site showed that the constraints on the group of piles also depended on their location below the raft. Reaction induced by raft bending had effects of the same order of magnitude as the mechanical loads carried by the individual piles.

2.5. Bibliography

[BOU 09] BOURNE-WEBB P.J., AMATYA B., SOGA K., *et al.*, "Energy pile test at Lambeth College, London: geotechnical and thermodynamic aspects of pile response to heat cycles", *Géotechnique*, vol. 59, no. 3, pp. 237–248, 2009.

[DUP 13] DUPRAY F., LALOUI L., KAZANGBA A., "Numerical analysis of seasonal heat storage in an energy pile foundation", *Computers and Geotechnics*, (in revision), 2013.

[GLI 00] GLIŠIĆ B., SIMON N., "Monitoring of concrete at very early age using stiff SOFO sensor", *Cement and Concrete Composites*, vol. 22, no. 2, pp. 115–119, 2000.

[INA 00] INAUDI D., LALOUI L., STEINMANN G., "Looking below the surface", *Concrete Engineering International*, vol. 4, no. 3, 2000.

[KNE 11] KNELLWOLF C., PÉRON H., LALOUI L., "Geotechnical analysis of heat exchanger piles", *Journal of Geotechnical and Geoenvironmental Engineering*, vol. 137, no. 10, pp. 890–902, 2011.

[LAL 99] LALOUI L., MORENI M., STEINMANN G., *et al.*, Test en conditions réelles du comportement statique d'un pieu soumis à des sollicitations thermo-mécaniques, Swiss Federal Office of Energy (OFEN) report, 1999.

[LAL 03] LALOUI L., MORENI M., VULLIEt L., "Comportement d'un pieu bi-fonction, fondation et échangeur de chaleur", *Canadian Geotechnical Journal*, vol. 40, no. 2, pp. 388–402, 2003.

[LLO 00] LLORET S., INAUDI D., GLIŠIĆ B., *et al.*, "Optical set-up development for the monitoring of structural dynamic behavior using SOFO sensors", *Smart Structures and Materials 2000: Sensory Phenomena and Measurement Instrumentation for Smart Structures and Materials*, vol. 3986, Spie-Int Soc Optical Engineering, pp. 199–205, 2000.

[MAT 08] MATTSSON N., STEINMANN G., LALOUI L., "Advanced compact device for the *in situ* determination of geothermal characteristics of soils", *Energy and Buildings*, vol. 40 no. 7, pp. 1344–1352, 2008.

[MIK 02] MIKKELSEN P.M., "Cement-bentonite grout backfill for borehole instruments", *Geotechnical Instrumentation News*, pp. 38–42, December 2002.

[STR 91] STRAIN R.T., WILLIAMS H.T., "Interpretation of the sonic coring results: a research project", *Proceedings of the 4th International Conference on Piling and Deep Foundations*, pp. 633–640, 1991.

Chapter 3

Observed Response of Energy Geostructures

Dating from the mid-1990s until the present day, a number of publications have presented performance data relating to the use of energy geostructures, primarily pile foundations. These have included observations of either the thermal performance only or the combined thermomechanical performance, and the observations have come from operational systems, field trials and, more recently, laboratory experiments. This chapter brings many of these publications together in one place and aims to provide an overview of the information that is currently available and that addresses the issues associated with the performance of energy foundations.

3.1. Overview of published observational data sources

While not claiming to be 100% comprehensive, the information summarized in the tables that follow at the end of this chapter provide a fairly comprehensive overview of the information that is available in the public domain. Given that this application originated in Switzerland, Austria and Germany, there is a surprising lack of detailed case studies from these countries that can be included in this review.

The information collated here has been arranged as follows: each pair of tables summarizes the details for operational systems with thermal observations (Table 3.4), field tests with ground and geostructure temperature observations (Table 3.5), field tests with geostructure temperature observations only

---

Chapter written by Peter BOURNE-WEBB.

(Table 3.6) and cases where observations have been made of the thermal and mechanical response of the geostructure (Table 3.7).

The first table in each pair (Tables 3.4a, 3.5a, 3.6a and 3.7a) summarizes the location, ground conditions (where available), and energy geostructure details (element type, type of construction, cross-section dimension, length, spacing and number). A wide range of ground conditions is reported, however, in all but two cases [NAG 05, SEK 05]; the soils can be considered to be saturated with a groundwater level within about 3 m of the ground surface. Only one case [NAG 05] reported any significant groundwater flow.

The second table (Tables 3.4b, 3.5b, 3.6b and 3.7b) summarizes some of the reported details relating to the thermal performance of the energy geostructure (inflow temperature, flow rates, power output, thermal conductivity and coefficient of performance (COP/SCOP)).

A total of 20 cases from seven different countries have been collected and of these 18 relate to tests on various types of piles; 1 case is for a plain raft foundation used for thermal exchange [SCH 06, KIP 09] and finally, 1 case involves a set of thermal response tests in a diaphragm wall [XIA 12]. [BRA 06] refers to the use of piled and diaphragm walls without quoting operational performance data.

Reported operational and field test data relate to a range of pile types (Figure 3.1), including bored cast *in situ* [SEK 05, BRA 06, LAL 06, PAH 06, SCH 06, GAO 08, BOU 09, KIP 09], auger cast *in situ* (continuous flight auger) [WOO 09, BRE 10], driven or screwed steel tubes that were either grout/sand in-filled [HEN 98, JAL 11] or water filled [MOR 94, NAG 05, KAT 09, JAL 11] and precast concrete piles that were either circular hollow sections [HAM 07, KAT 09] or square [BRA 06]. Chapter 2 of this book also presents information relating to the field testing of piled energy foundations.

3.2. Thermal storage and harvesting

3.2.1. *Overview*

All of the cases provide some information relating to the thermal response of the energy geostructure, and of these, seven also reported observations of the temperature field in the ground surrounding the structure [NAG 05, SEK 05, BOU 09, WOO 10, FRA 11, SCH 12, JAL 11], see section 3.2.3.

Typically, the reported observations included inflow and outflow temperatures at the geostructure, circuit fluid flow rates and power consumption. In addition to these measurements, many of the cases also report temperatures at various locations within the geostructure, and in the cases mentioned above temperatures were observed in the ground, at various depths and distances from the geostructure.

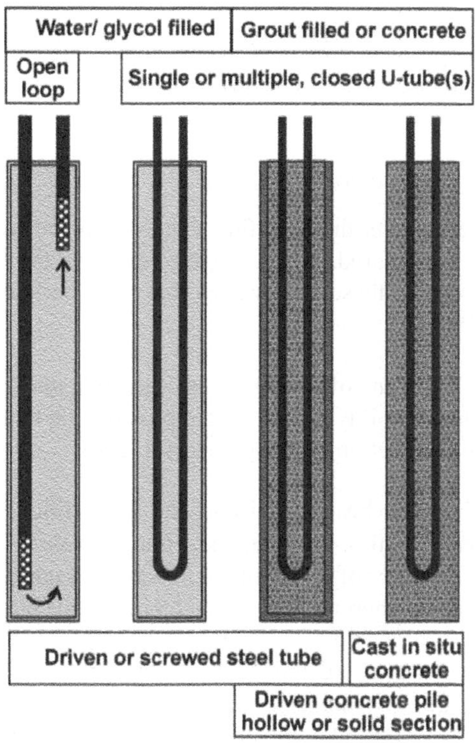

Figure 3.1. *Summary of energy pile types and methods of heat exchange*

3.2.2. *Energy injection/extraction rates*

It seems that, traditionally, the heat transfer (flux) in or out of the borehole heat exchangers (BHX) has been reported in terms of Watts per meter (W/m), and this approach has also been adopted in most of the case studies presented here. This is presumably because this is a familiar way in which the performance of geo-heat exchanger (GHX) systems can be compared, rather than for its scientific merits.

As a result, this measure is also used here where it has been reported or can be evaluated from the published information. However, the values for heat exchange from different cases are very difficult to compare because of variations in ground and groundwater conditions, pile type, pile dimensions, pile spacing, test methodology, thermal loading characteristics and whether a steady-state condition was approached. Short heating or cooling cycles [e.g. MOR 94, SEK 05, HAM 07, KAT 09], while perhaps more representative of operational conditions, will yield higher values of heat flux than longer, semi-steady-state tests [see NAG 05, GAO 08, BRE 10, JAL 11].

In addition, where possible, the thermal conductivity of the pile-soil system and coefficient of performance (COP) are also presented. In some cases, a seasonal coefficient of performance (SCOP) is quoted.

3.2.2.1. *Piled energy geostructures*

Although, as noted above, the heat flux values obtained in the cases reviewed here cannot be readily compared, it is possible to examine the results for any broad trends and also to compare these with values that have been suggested for design, see [BRA 06] and [BS 07].

The results have first been collated in terms of the pile material (concrete or steel tube with water infill) and soil type, and where possible the reported values for heat injection (cooling) and extraction (heating); see Table 3.1.

Although the volumetric heat capacity of the water-filled steel piles is almost double that of the concrete piles, the actual heat transfer values barely differ because the steel piles have a diameter of less than 0.4 m while the concrete piles are 0.6–1.5 m in diameter, and thus the thermal mass of each pile type is broadly similar.

The results tend to cover a large range of values; however, there does seem to be a trend of higher heat transfer in coarse-grained soil profiles than those that are fine grained, with fluid-filled steel tube piles possibly yielding higher heat transfer than concrete piles.

The comparison between the values suggested for borehole installations by BS 15450:2007 [BSE 07] and those in Table 3.1 is interesting. Good agreement is seen for the steel pile cases where the ratio of pile length to pile diameter, L/D, is in the range of 30 to 240 but the Code values appear to be somewhat un-conservative for concrete piles where the L/D ratio lies in the range of 10 to 60.

The BS 14450 values are intended for the pre-design of BHX, where the L/D ratio is likely to be in the range of maybe 200–500 (i.e. 0.15–0.30 m diameter boreholes and 80–150 m deep). Thus, as the pile aspect ratio approaches these values it is reasonable that the recommended values become more appropriate.

Material	Soil type	Heat transfer, W/m		BS 15450 values by soil type
		Injection	Extraction	
Concrete	Fine	15–60, max. 110	25–45	30–50
	Coarse	35, max. 220	30–50	55–80
Steel tube with fluid infill	Fine	25–55, max. 140	15–20, max. 85	30–50
	Coarse	55–90	–	55–80

Table 3.1. *Comparison of heat transfer by pile material type*

A number of the reported results in Table 3.1 are very high and Table 3.2 seeks to explain these by collating those cases where the heating was applied either in pulses, that is heat injection or extraction over varying time periods (typically up to 12 h), or until such time that a steady-state response was approached.

Thermal loading	Heat transfer, W/m	
	Injection	Extraction
Intermittent	70–120, max. 220	50–85
Steady state	25–60, max. 110	15–25

Table 3.2. *Comparison of heat transfer by thermal loading type (excluding operational systems)*

It is clear that heat pulses (or intermittent heat transfer) result in a higher heat transfer value than that in the steady state. [JAL 11] report heat transfer values throughout a test to the steady state which illustrate this point – values of 125 W/m, 65 W/m and 35 W/m were obtained for a double U-tube in a 0.140 m diameter pile, after 1, 4 and 24 h, respectively.

The comparison between continuous and intermittent operations is further illustrated in Figure 3.2 [JAL 10] where the heat transfer from differing heat-exchanger tube configurations during the last 2 h of a test lasting 16 h (and approaching a steady state) is compared with a test of less than 2 h duration. The higher average transfer value during intermittent operation is clearly apparent.

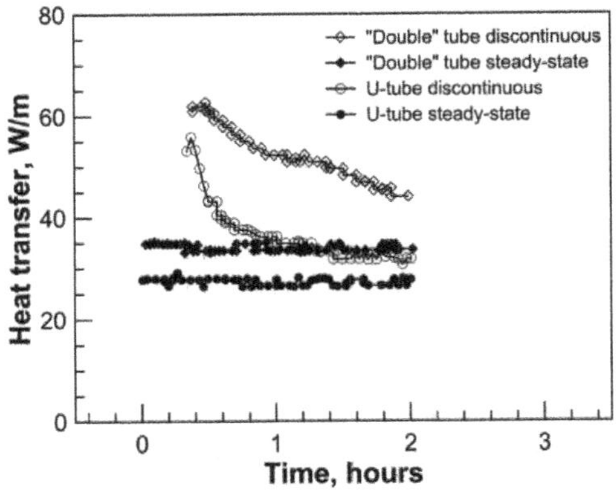

Figure 3.2. *Effect of intermittent versus continuous heat transfer, after [JAL 10]*

The intermittent heat transfer cases are perhaps more representative of an operational system that might only provide heating or cooling for part of the day, and maybe only 5 or 6 days per week. Thus, the results from TRT that achieve a steady state should provide a safe estimate of the actual thermal performance of an operational system.

Referring back to Table 3.1, the heat transfer values suggested by BS EN 15450:2007 [BS 07] (30–80 W/m depending on the soil type) may in fact be suitable approximations for pile thermal exchange systems that are operated intermittently. However, this approach requires verification to ensure the comparison is made on a like-for-like basis in terms of intermittent thermal loading of piles.

Finally, the results can be compared in terms of diameter of the pile and with values suggested by [BRA 06] for use in pre-design of thermal pile systems (Table 3.3). For small diameter piles, the range of values suggested by [BRA 06] lies within the range of results obtained in the case studies, it is apparent that in some situations these values may not be conservative and in others they could be excessively so. However, they are likely to be suitable as a first approximation in order to provide an initial layout for use in a more refined analysis. Again, this perhaps reflects that small diameter piles tend to have larger L/D ratios.

There were only four cases with pile diameters larger than 0.6 m, however, there was no obvious relationship between pile diameter and heat transfer. [BRA 06] recommends using a value of 35 W/m^2 for preliminary design of piles greater than 0.6 m diameter; [GAO 08] quote (steady-state) values in the range of 30–60 W/m^2 for a 0.6 m diameter pile while [SEK 05] quote 25–45 W/m^2 for a 1.5 m diameter pile under pseudo-operational intermittent conditions which suggests that this may be a reasonable first estimate for an outline design.

Pile diameter	L/D	Heat transfer	Pre-design values [BRA 06]
<0.5 m	30–240	20–90 W/m, max. 140 W/m	40–60 W/m
>0.6 m	10–60	20–60 W/m^2 (injection)	35 W/m^2
		15–25 W/m^2 (extraction)	

Table 3.3. *Comparison of heat transfer by pile diameter*

[LOV 12] highlight the key differences between the likely thermal performance of pile foundations and BHX, primarily as a function of the large thermal mass of large diameter piles and the length to diameter ratio of piles when compared to BHX, and how, as a result, the use of existing line-source models (used for design of BHX) may lead to significant errors in the prediction of the thermal response of piles. [LOV 12] conclude that there is a pressing need for high-quality observational data sets to allow this issue to be understood and to support the development of new thermal analysis tools for the design of energy foundations.

3.2.2.2. *Other energy geostructures*

While there is a reasonable body of thermal performance information for piles, very little exists for other types of energy geostructure such as rafts/ground slabs, piled or diaphragm walls, or tunnels.

A study examining the efficiency of differing heat-exchanger tube configurations in a diaphragm wall panel illustrates how the configuration and lateral spacing of the tubes affect the heat exchange rate [XIA 12] (Table 3.6a). At least the relative effects could be studied; it appears that the test configuration considered only a single loop and not the effect of a series of loops along the wall. In a complete installation, the loops along the wall would interfere with each other and limit the amount of heat flow laterally in the concrete panel. Thus, the heat exchange rates (30–100 W/m length of exchanger tube) quoted in this study (Table 3.6b) may not be representative of the performance of an operational system. [BRA 06] suggests that a value of 30 W/m^2 could be used for thermal pre-design of diaphragm walls.

The Lainzer railway tunnel [BRA 06, ADA 09] featured a number of innovative energy-harvesting systems, including a section of bored pile cut-and-cover tunnel (LT24) and a section of sprayed concrete lined (SCL) tunnel (LT22). Results from the latter have not been published, however, some results from LT24 have been presented. [ADA 09] presents a plot of cumulative heat extraction for the first 2.5 years of operation of the system, and over the second and third heating seasons the total heat extracted averaged about 190 MWh per season (each lasting 8–9 months); this equates to heat extraction of about 30 W/m length of pile and a COP of about 3.

The Fasanenh of tunnel is a SCL-type tunnel, parts of which have been thermally activated. Thermal observations from one of these installations are reported by [SCH 12] (Figure 3.3). This illustrates that, over a period of 4 months as the outside air and tunnel air temperatures rise, heat transfer falls from about 30 W/m^2 to 15 W/m^2 (the authors indicate that the theoretical steady-state value is between about 5 W/m^2 and 8 W/m^2).

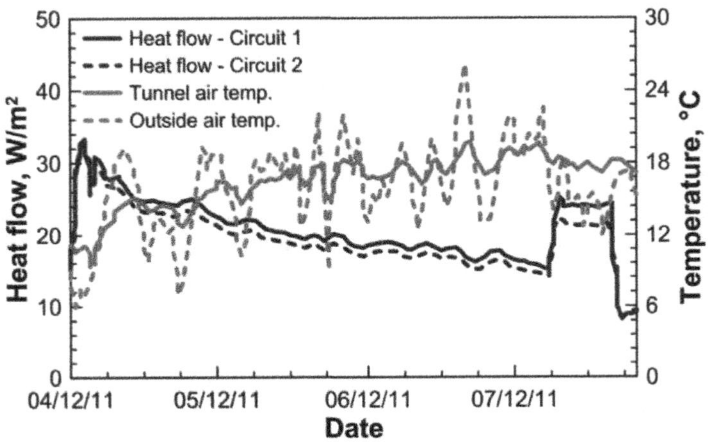

Figure 3.3. *Heat transfer from tunnel to rock mass, from [SCH 12]*

The thermal performance of a thermally activated 8,000 m^2 raft in Berlin was monitored for a period of 4 years and is reported as yielding about 5 W/m^2 in heating and cooling [SCH 06, KIP 09]. However, it is unclear how much this value might have been compromised by the operational problems associated with the free cooling mode, identified in the papers. As these issues were addressed, the COP rose from about 2.5 in 2005 to 5.0 in 2008, with an expectation that this should reach a value of at least 10 across the year. In comparison, [BRA 06] suggests that a value of 10–30 W/m^2 could be used for pre-design of thermally activated base slabs.

3.2.3. *Thermal fields*

Of the 20 case studies presented here, seven (five pile/pile groups and two tunnels) include measurements of the temperature field in the ground adjacent to the energy geostructure. The results of these observations are summarized below.

3.2.3.1. *Pile internal temperatures*

Although, in many cases, instrumentation was said to have been provided for observations of the pile temperature, only a few cases have actually presented the results [MOR 94, NAG 05, HAM 07, JAL 11, BRA 06, LAL 06, BOU 09, AMA 12].

Typically, the temperature of the pile body has been found to be largely uniform while at the upper and lower extremities of the pile the temperatures differ somewhat. For example, in Figure 3.4 [NAG 05], while the temperatures in the body of the pile cool, the near-surface temperature increases throughout the test period, in response to increasing air temperatures moving from late-Winter into Spring. Meanwhile, at the pile tip, cooling is evidenced but is not as strong due to thermal recharge from the soil mass below.

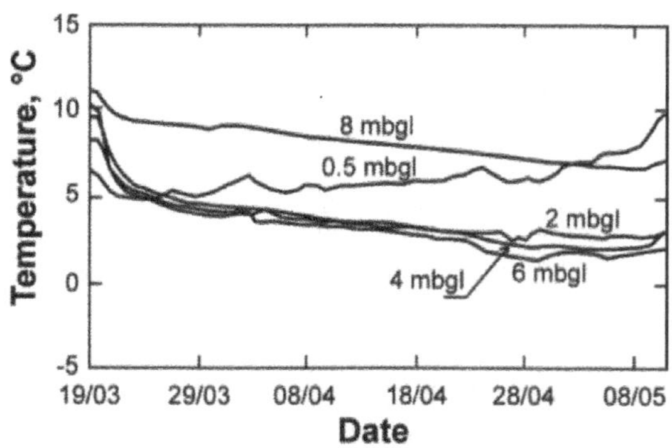

Figure 3.4. *Surface temperature of pile at center of group, after [NAG 05]*

Data from [BOU 09] and [AMA 12] also suggest a similar depth of influence with observed pile temperatures being effectively constant below a depth of 4–6 m below ground level (bgl). In the case of the Lambeth College test pile [AMI 08, BOU 09], temperature readings from different instruments (thermistors and optical fiber sensors, OFSs) at the same level within the pile suggested some variation in the pile section; the thermistors located mid-way between heat-exchanger tubes generally registered slightly higher temperatures than the OFSs that were located closer to the tubes.

The temperature variation along the length of the heat-exchanger tube within a pile during heat rejection to the ground has been measured by [JAL 11]. As would be expected, the temperatures fall with depth down the pile and recover somewhat as they return, presumably due to interference between the supply and return pipes.

The overall steady-state change in temperature, ΔT (inflow minus outflow temperature), seen in these experiments, is reported to be 1°C–4°C (increasing with reducing flow rate). However, much of this change appears to occur in the near surface region, between the upper most set of gages in the pile and the point where the inflow and outflow temperatures were taken.

At the upper most set of temperature readings in the pile, about 2 m depth, the apparent difference between inflow and outflow temperatures reduces to values generally less than about 2°C. This suggests that the outcome of the test, in terms of the heat transfer reported, was strongly influenced by conditions at the pile head; air temperatures (consistently cooler than both the inflow and outflow temperatures during the test), where the inflow and outflow temperatures were observed and how well the exposed parts of the system were insulated, will all have played a role in this effect.

This effect may be acceptable for a borehole system that will be exposed to the open-air when operational. However, it would seem that some care must be taken to ensure that the thermal performance obtained in a test is consistent with the thermal boundary conditions that will apply in an operational energy geostructure. For example, pile heads are not exposed to atmospheric conditions but are insulated by the overlying building that they are needed to support.

3.2.3.2. *Ground temperatures around single piles*

The temperature field in the ground around a single energy pile was observed and reported by [SEK 05], [BOU 09] and [JAL 11].

The seasonal variation in undisturbed ground temperature at different depths in the ground profile is illustrated in Figure 3.5, after [JAL11], and highlights that the seasonal climatic variation in air temperature is significant to a depth of about 4 m.

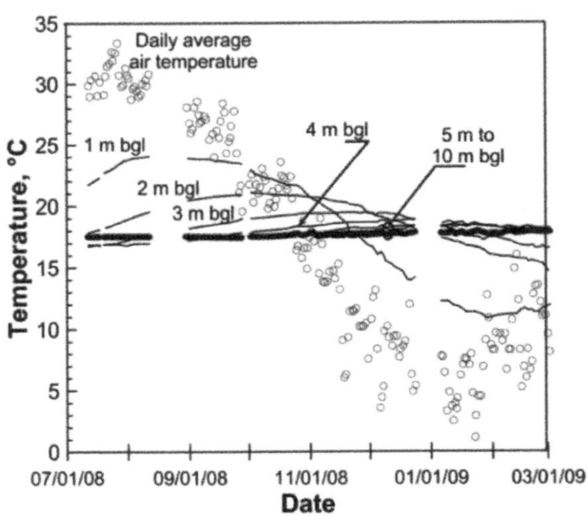

Figure 3.5. *Seasonal ground temperature variation, after [JAL 11]*

Data from [BOU 09] also suggest a similar depth of influence with observed pile temperatures being effectively constant below a depth of 4–6 m below ground level The temperature data in this case are complicated by the presence of a heat source in the form of a metro tunnel approximately 25 m from the test pile.

The observations in Figure 3.6 [SEK 05] illustrate the typical spatial variation around a pile that is being heated (cooling season) and cooled (heating season). Figure 3.6(a) represents observations at a distance of 0.67R (R = radius) from the pile face and Figure 3.6(b) 1.67R.

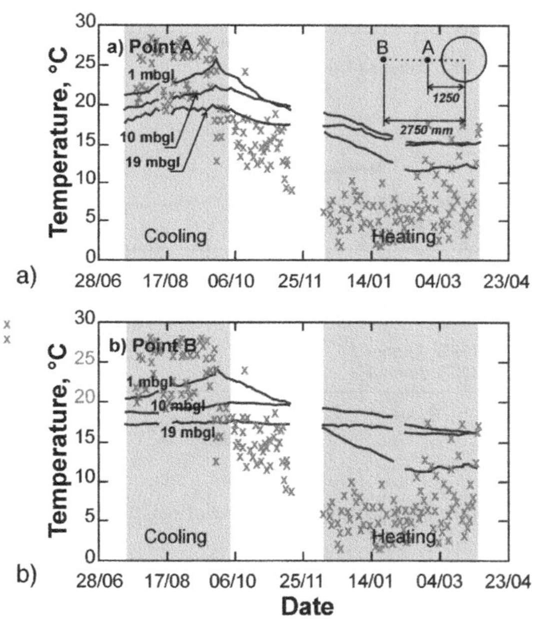

Figure 3.6. *Ground and air temperature adjacent to energy pile, after [SEK 05]*

Near the surface, the ground temperature is affected by both pile and climatic air temperature variations and shows more variation (+10°C to –5°C with respect to an undisturbed ground temperature of about 15°C) than the observations at depth (+7°C to 0 with respect to undisturbed). Temperatures in the ground at the level of the pile toe show the least variation and this is most probably due to thermal recharge from the soil mass below.

The observations in Figure 3.6 also illustrate the reduction in temperature change that occurs with increasing distance from the pile, as do the results from [BOU 09] in Figure 3.7 that are from a level of 9.5 m below ground level. Temperature in the borehole at a distance from the pile face of 0.83R was about half the values measured in the pile, while at a distance of about 5.67R there was negligible temperature change observed in the anchor piles.

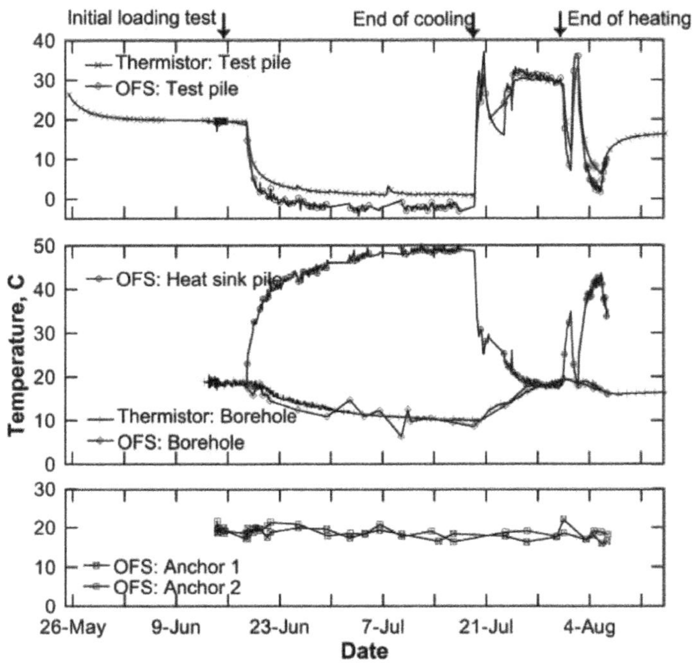

Figure 3.7. *Variation in ground temperature adjacent to energy pile, after [BOU 09]*

3.2.3.3. *Ground temperatures around pile groups*

Observations from experiments involving groups of piles are reported by [NAG 05] and [WOO 10]. These data sets illustrate the interaction that might occur between adjacent piles when they are thermally loaded.

In the former case, the piles were spaced in a 5×5 grid at a center-to-center spacing of about 24R [NAG 05]. In the latter, only the peripheral piles (16 in number) of a group of 21 piles were thermally loaded, and these were installed at a spacing of about 12–16R [WOO 10].

Temperature observations from [NAG 05] are presented in Figure 3.8 where the areal distribution of temperature on a plane 6 m below ground level is examined. The effect on ground temperature moving from a remote location (point 8) to the location near or on the edge of the pile group (points 1, 5, 3) and onwards toward the center (points 2, 6, 4, 7, 9) is clear.

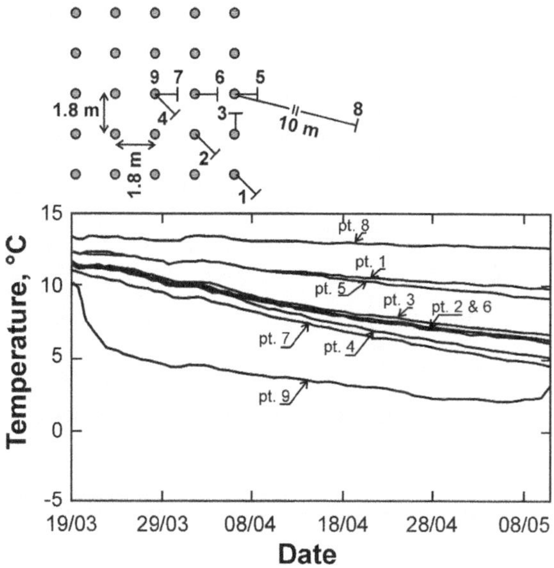

Figure 3.8. *Ground temperature adjacent and within a pile group, after [NAG 05]*

At a distance of 10 m (138R) from the edge of the group (point 8) the temperature is largely unaffected (there is possibly a very small cooling effect by the end of the test), at a distance of 12–18R (points 5 and 1, respectively) cooling increases to 3°C–4°C, while at the edge (point 3) the effect increases by another 1°C–2°C, and further increases are apparent toward the center of the group.

This illustrates the superposition of thermal fields from the adjacent piles. Similar effects are reported by [WOO 09] and [WOO 10] when examining the temperature field within a pile group subject to heat extraction only. They also showed that at a distance of about 32R from the corner of the group, the observed temperature variation along a vertical profile was due to the variations in surface air temperature only. At half of this distance, the effect was less than 1°C cooling.

3.2.3.4. *Other energy geostructures*

Temperature fields around energy geostructures other than piles are very rare with the only cases found being for tunnels [PRA 09, FRA 11, SCH 12]. Reporting observations from the segmental lined Jenbach tunnel [FRA 11] illustrate the distance that thermal loading in the tunnel lining penetrates the surrounding ground (Figure 3.9). Similar observations from an SCL tunnel but in terms of how natural air temperature variation in the tunnel penetrates the surrounding ground were reported by [SCH 12].

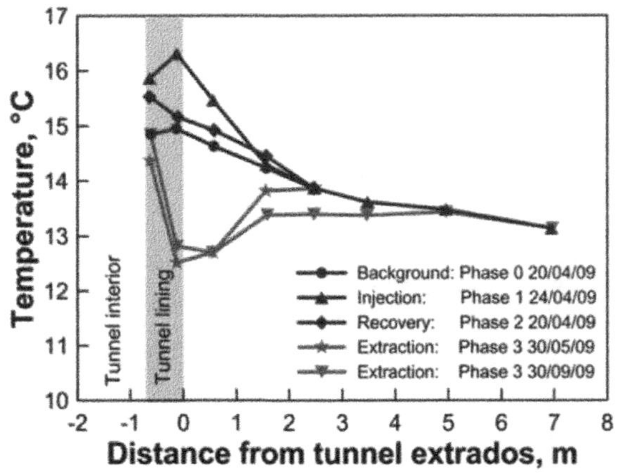

Figure 3.9. *Observed variation in tunnel lining and ground temperature with distance from tunnel extrados, after [FRA 11]*

In both data sets, the influence of temperature variation whether natural or in the former case, forced by the injection and extraction of heat, penetrated no more than about 7–8 m from the tunnel extrados (a little less than one tunnel diameter).

3.3. Thermomechanical effects

3.3.1. *Overview*

Published observations of the impact of thermal loads on the mechanical response of full-scale energy geostructures are currently limited to piled foundations [BRA 06, LAL 06, BOU 09, AMA 12] and piled walls [BRA 06]. In addition to the full-scale energy geostructure systems described above, a number of small scale 1g and centrifuge model experiments have also been executed [MCC 11, KAL 12, STE 12, WAN 12]. See also, Chapters 3 and 4 of this book.

3.3.2. *Structural effects*

A detailed examination of the reported effect of heating and cooling on the observed response of full-scale piles was made by [AMA 12]. In essence, they confirmed that the effects seen were a function of the pile characteristics (geometry, Young's modulus, etc.), the thermal load applied (change in temperature) and the level of restraint applied to the pile (by the surrounding ground and from structures supported by the pile).

Cooling the pile (building heating season) led to the stresses in the pile section becoming less compressive with respect to the initial stresses due to external mechanical loading, and given sufficient cooling the stresses in a compression pile could even become tensile (Figure 3.10(a-i).

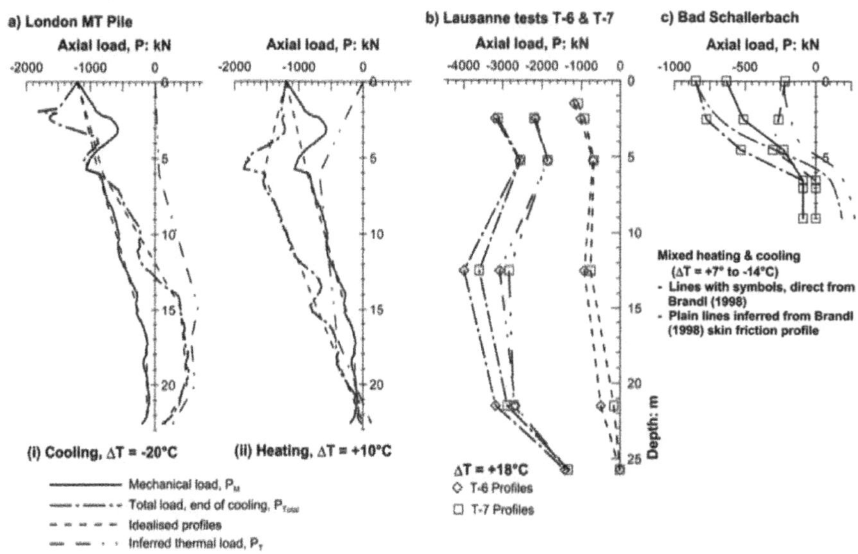

Figure 3.10. *Change in axial stress in piles due to heating and cooling, after [AMA 12, BOU 09, LAL 06, BRA 06]*

Heating, on the other hand, resulted in increased compressive stress, the magnitude of which, in part, depends on the degree of freedom for movement at the pile head. In Figure 3.10(a-ii), the pile head is free to move, while in Figure 3.10(b) the pile was effectively fixed between a stiff overlying structure at the pile head and stiff sandstone at the pile toe. A similar effect in response to restraint at the head and toe of the pile was reported by [STE 12].

Examining these axial stress changes and comparing them with the maximum values that could be generated if the pile was perfectly fixed against thermal movement (equation [3.1]) yields some interesting insights (Figure 3.11):

$$\sigma_{\text{a,fixed}} = \alpha\,\Delta T.E_p \qquad [3.1]$$

where $\sigma_{\text{a,fixed}}$ is the pile axial stress when fully restrained against thermal movement, α is the coefficient of linear thermal expansion, ΔT is the change in temperature and E_p is the Young's modulus of the pile section.

The maximum thermally induced axial stress changes observed in the piles ranged between about 50% and 100% of the fully restrained value. It is apparent and logical that the fully restrained value is an upper bound to the observed changes in stress. Further research is needed to understand the mechanisms of restraint and how to quantify intermediate levels of restraint, that is cases of less than perfect restraint as seen in the cooling phase in Figure 3.11(a) and during heating in Figure 3.11(b).

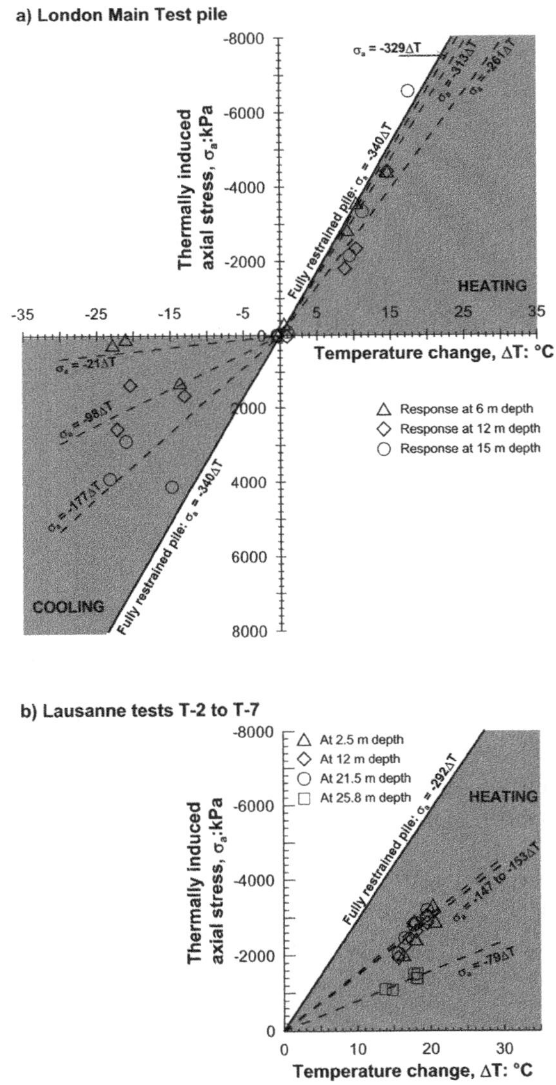

Figure 3.11. *Mobilized concrete stress in pile shaft relative to theoretical fully restrained value, after [AMA 12]*

The only other observation of thermomechanical response of a full-scale energy geostructure that has been published is that for the Lainzer tunnel in Austria where the impact of temperature changes on the bending response of a secant piled wall in a cut-and-cover tunnel are presented [BRA 06]. The presentation of the data from [BRA 06] in Figure 3.12 has been modified from the original paper to show the changes in observed strain during differing periods rather than absolute values. As in [BRA 06], the data include both mechanical bending strain and thermally induced strain.

Between March 2003 and August 2003, it is not apparent from [BRA 06] what was actually happening to the wall system either physically or thermally. However, in absolute terms the strain profiles reported by [BRA 06] have the shape that would be expected for a wall propped near its top, after completion of excavation. This means that the largest bending strains in the wall occur just above the final excavation level. The changes in strain that occur in this period (Figure 3.12(a)) imply that all the strains become more tensile, though not uniformly. In the same period, air temperatures will have increased from near zero to the mid- to high-twenties, so the tensile changes in strain can be explained in part by the expansion of the wall piles in response to seasonal warming.

The outer side (exposed to the tunnel air void) was warmer [BRA 06], and thus the tensile change in strain is larger than on the side against the soil. It is difficult to explain the total difference via this mechanism, however, when the reported temperature differential across the pile section was less than 3°C [BRA 06] and perhaps physical restraints within the wall system and/or construction activity not noted by [BRA 06] contributed to the remaining strain changes. The maximum change in strain is about 200 με which is about 40% of the maximum absolute strain recorded in March 2003.

Similarly, between August 2003 and January 2004, and again prior to heat-pump commissioning, Figure 3.12(b), the outside air temperature and hence the tunnel air temperature falls. As a consequence of this cooling, the piles contract and the recorded change in strain is compressive and of a similar magnitude to the maximum in the previous period. In this instance, however, in the zone above the base slab, the observed changes in strain are more uniform both at each level of gages and along the length of the wall.

In August 2004, the air temperatures were a few degrees cooler than those in the previous August and it can be seen (Figure 3.12(c)) that the observed change in strain at this point was still compressive but of a much reduced magnitude as would be expected. Observations from the period August 2003 to June 2004 reported by [BRA 06] are not shown here but have a very similar form to that in Figure 3.12(c), with strain magnitudes somewhat higher due to the cooling relative to the previous August (about 10°C) being between that of the periods to January 2004 and August of 2004.

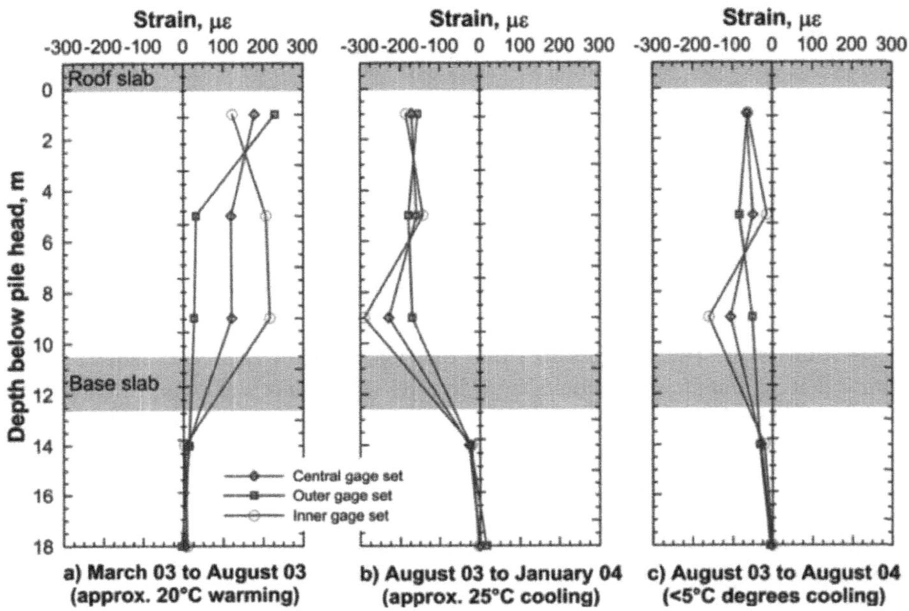

Figure 3.12. *Changes in strain in secant pile wall during a cooling season, after [BRA 06]*

By August 2004, the ground source energy system had been in operation since the start of February 2004 and it is clear that decoupling the effect of heat extraction/injection from the climatic changes in the temperature and strain data is problematic. However, and as was noted by [BRA 06], when comparing the pre-operational period with the post, it is clearly apparent that the air temperatures in the tunnel seem to dominate the observed response – or rather, maybe the heating and cooling demands are such that thermal exchange with the tunnel air volume is sufficient to meet the energy demand. The relatively unchanging strain response below base slab level, seen in each of the three reference periods illustrated in Figure 3.12, perhaps supports this.

3.3.3. *Soil-structure interactions*

The effect of heating and cooling on the stresses acting at the soil-structure interface has been published for piles only, and the available full-scale observations were collated and examined by [AMA 12] (Figure 3.13).

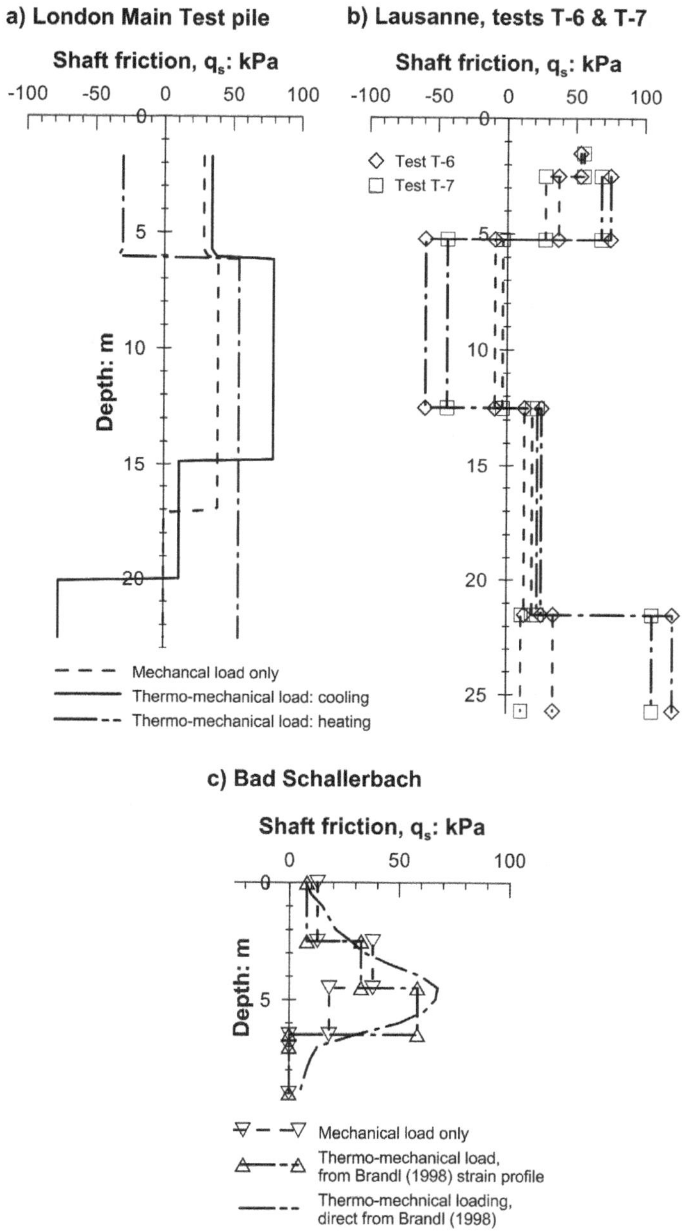

Figure 3.13. *Alteration of the patterns of shaft friction mobilization due to heating and cooling of piles, after [AMA 12]*

In the figures, under vertical mechanical loading only, the soil opposes pile movement into the ground, and the soil-pile interface friction is plotted as a positive value. Friction mobilized at the soil-pile interface changes in response to the thermal expansion/contraction of the pile. Thus, when the pile is cooled, the mobilized friction increases in the upper parts of the pile (the pile is contracting down into the soil) and reduces, becoming negative in the lower part (the pile is contracting up out of the soil), solid line in Figure 3.13(a). Observing the figure, the reverse happens during heating.

Model pile tests are being used to examine the effect of thermal loading on the available friction resistance at the pile-soil interface. In-line with the findings of the full-scale tests, after testing a model pile in dry sand at 1g, [KAL 12] found that at mechanical load levels representative of a typical design factor-of-safety (global factor of about 2.5) the evolution of pile head movement as the pile was heated and cooled (ΔT = 25°C) was elastic, that is no permanent pile settlement occurred. However, as the mechanical load was increased toward the ultimate pile resistance, significant additional settlement accumulated following cycles of heating and cooling at constant load (noting that there would have been a component of creep in the pile system that the authors had not decoupled from the observations) [KAL 12]. [WAN 12] carried out 1g model pile-soil interface tests to examine the effect of temperature on available shaft friction. The tests were undertaken using compacted uniform fine sand (D_{50} = 0.21 mm; γ_d = 15.2 kN/m^3), at moisture contents of 0%, 2% and 4%. Each test involved several stages as follows:

1) cyclic loading to establish a steady-state resistance at room temperature;

2) apply change in temperature ΔT = 20°C or 40°C for 24 h;

3) cyclic loading to establish a steady-state resistance at elevated temperature;

4) allow model to cool at room temperature for 24 h;

5) cyclic loading to establish a steady-state resistance at room temperature.

Repeating steps (2)–(5) as necessary.

Accumulating settlement at effectively constant resistance was observed in dry sand and sand at 4% moisture content with ΔT of 20°C and 40°C, and in sand at 2% moisture content with ΔT of 20°C. Noting that the available resistance in sand with 2% and 4% moisture content was approximately twice that of dry sand.

With a soil moisture content of 2%, a temperature increase ΔT of 40°C resulted in the load resistance halving, effectively to that of dry sand. A repeat test carried out under the same conditions not only captured the same effect, but also showed that the available resistance recovered when the pile cooled back to its initial temperature. Soil moisture content determinations adjacent to the pile shaft were revealing; in the case where a large drop in resistance was observed, the moisture content had reduced to about 0.2% while in all other cases the values remained above 1%, including the second test where the temperature was allowed to recover [WAN 12].

It would seem that moisture movement in the soil driven by the thermal gradients may have contributed to the observed changes in shaft resistance, that is the soil adjacent to the pile shaft has effectively dried and suctions that contributed to higher resistance (at higher moisture content) have broken down, with the available resistance dropping to that of dry sand. In this instance, this appears to be a temporary effect because when allowed to cool, the available shaft resistance recovered, that is soil moisture perhaps migrated back toward the shaft and suctions were re-established.

Testing partially saturated silt with a moisture content of about 13% [MCC 11] found that a 40°C temperature increase led to an increase in pile resistance of about 30%. This effect could also be attributed to the effect of temperature in altering the soil suction, i.e. increasing temperature leads to increasing curvature of the water meniscus and higher suction, and thus perhaps higher pile resistance.

3.4. Summary

Various values for heat transfer between energy geostructures (primarily piles) and the ground have been quoted in the literature either as test-specific results or as ranges that could be used for pre-design (see [BRA 06]). However, these values should be treated with a great deal of caution as the specific details of an installation (ground conditions, size(s) and layout of the geostructure, thermal boundary conditions, heating and cooling demand profile) will have a major impact on the heat transfer characteristics of the energy geostructure. This is clearly demonstrated by the range of results that are quoted. Ultimately, only project-specific testing and analysis will yield an optimal energy design for any geostructure.

The zones in which there are significant fluctuations in temperature fields around energy geostructures appear to extend no more than a few meters in the case of piles, and possibly pile groups to several meters in the case of tunnels. As is the case with influence zones associated with mechanical loading, the factors that define the extent of the influence zones around energy geostructures need to be better understood, in order for thermal interference between adjacent ground energy systems to be minimized. Again, only project-specific testing and analysis will define the thermal influence zone of a particular energy geostructure.

From what is at the present time, a rather limited database, it would appear that the impact of thermal loading on piles in terms of movement and internal stress can be bounded based on the theoretical limits defined by a perfectly free and perfectly fixed column respectively [AMA 12]. However, in many cases, this could prove to be excessively conservative and so further observation and analysis are needed in order to fully appreciate the key parameters affecting the pile response, and the sensitivity of the response to variations in these parameters.

The impact of temperature and temperature cycles at the soil–structure interface has been observed on piles at full and model scales, and it appears that the changes in mobilized friction are reversible when applied mechanical loads have a conventional margin of safety with respect to the pile ultimate load resistance. Model-scale tests have also highlighted the important impact that soil moisture migration, induced by thermal gradients, might have on available pile resistance, especially in partially saturated soils. Here again, key variables need to be identified and their effect observed and analyzed in order to fully understand the impact of heating and cooling on energy geostructures.

While the above discussion relates primarily to observations and our current understanding of the response of pile foundations, there are no equivalent groups of studies (yet published) that evaluate these responses for other types of energy geostructure in different conditions, and this lack needs to be addressed in order to provide observational evidence of the adequacy of these systems.

Research on this subject is highly dynamic and the understanding of the response of energy geostructures is advancing rapidly. In the near future, the observations presented here will be supplemented by a number of investigation programs in a variety of countries which will help to provide a more complete picture of the thermal and thermomechanical response of energy geostructures, and greater certainty as to the engineering performance of the geostructures.

3.5. Acknowledgments

Many thanks are extended to Prof. Nagano (Hokkaido University), Dr. Chris Wood (University of Nottingham), Tony Amis and Chris Davidson (GI Energy), Prof. Daniel Pahud (SUPSI) and Hugh Henderson (CDH Energy Corp.) who all took time out of their busy schedules to answer the author's questions.

3.6. Bibliography

[ADA 09] ADAM D., "Tunnels and foundations as energy sources – practical applications in Austria", *Proceeding Conference on Deep Foundations on Bored and Auger Piles*, Ghent, Belgium, pp. 337–342, 2009.

[AMA 12] AMATYA B.L., SOGA K., BOURNE-WEBB P.J., *et al.*, "Thermo-mechanical behaviour of energy piles", *Géotechnique*, vol. 62, no. 6, pp. 503–519, 2012.

[AMI 08] AMIS T., BOURNE-WEBB P., DAVIDSON C., *et al.*, "An investigation into the effects of heating and cooling energy piles whilst under working load at Lambeth College, Clapham Common, UK", *Proceeding of the 33rd Annual and 11th International Conference of the Deep Foundations Institute*, New York, 2008.

[BOU 09] BOURNE-WEBB P.J., AMATYA B., SOGA K., *et al.*, "Energy pile test at Lambeth College, London: geotechnical and thermodynamic aspects of pile response to heat cycles", *Géotechnique*, vol. 59, no. 3, pp. 237–248, 2009.

[BRA 06] BRANDL H., "Energy foundations and other thermo-active ground structures", *Géotechnique*, vol. 56, no. 2, pp. 81–122, 2006.

[BRE 10] BRETTMANN T., AMIS T., KAPPS K., "Thermal conductivity analysis of geothermal energy piles", *Proceeding of the 11th International Conference on Geotechnical Challenges in Urban Regeneration*, DFI2010, London, 2010.

[BS 07] BS EN 15450:2007, "Heating systems in buildings – design of heat pump heating systems", *BSI*, p. 50, 2007.

[FRA 11] FRANZIUS J.N., PRALLE N., "Turning segmental tunnels into sources of renewable energy", *Proceedings of the ICE Civil Engineering*, vol. 164, no. 1, pp. 35–40, 2011.

[GAO 08a] GAO J., ZHANG X., LIU J., *et al.*, "Numerical and experimental assessment of thermal performance of vertical energy piles: an application", *Applied Energy*, vol. 85, no. 10, pp. 901–910, 2008.

[HAM 07] HAMADA Y., SAITOH H., NAKAMURA M., *et al.*, "Field performance of an energy pile system for space heating", *Energy and Buildings*, vol. 39, no. 5, pp. 517–524, 2007.

[HEN 98] HENDERSON H.I., CARLSON S.W., WALBERGER A.C., "North American monitoring of a hotel with room size GSHPs", *Proceeding of the IEA Room Size Heat Pump Conference*, Niagara Falls, Canada, 1998.

[JAL 10] JALALUDDIN J., MIYARA A., TSUBAKI K., *et al.*, "Thermal performances of three types of ground heat exchangers in short-time period of operation", *International Refrigeration and Air Conditioning Conference*, Paper 1123, 2010, available at http://docs.lib.purdue.edu/iracc/1123.

[JAL 11] JALALUDDIN J., MIYARA A., TSUBAKI K., *et al.*, "Experimental study of several types of ground heat exchanger using a steel pile foundation", *Renewable Energy*, vol. 36, no. 2, pp. 764–771, 2011.

[KAL 12] KALANTIDOU A., TANG A.M., PEREIRA J-M., *et al.*, "Preliminary study on the mechanical behaviour of heat exchanger pile in physical model", *Géotechnique*, vol. 62, no. 11, pp. 1047–1051, 2012.

[KAT 09] KATSURA T., NAKAMURA Y., OKAWADA T., *et al.*, "Field test on heat extraction or injection performance of energy piles and its application", *Proceeding of the 11th International Conference on Thermal Energy Storage, Stockholm, EFFSTOCK 2009*, Paper 143, 2009, available at http://intraweb.stockton.edu/eyos/energy_studies/content/docs/effstock09/Posters/146.pdf.

[KIP 09] KIPRY H., BOCKELMANN F., PLESSER S., *et al.*, "Evaluation and optimization of UTES systems of energy efficient office buildings (WKSP)", *Proceeding of the 11th International Conference on Thermal Energy Storage, Stockholm, EFFSTOCK 2009*, Paper 43, 2009, available at http://intraweb.stockton.edu/eyos/energy_studies/content/docs/effstock09/Session_6_1_Case_studies_residential_and_commercia_buildings/43.pdf.

[LAL 03] LALOUI L., MORENI M., VULLIET L., "Comportement d'un pieu bi-fonction, fondation et échangeur de chaleur", *Canadian Geotechnical Journal*, vol. 40, no. 2, pp. 388–402, 2003.

[LAL 06] LALOUI L., NUTH M., VULLIET L., "Experimental and numerical investigations of the behaviour of a heat exchanger pile", *International Journal for Numerical and Analytical Methods in Geomechanics*, vol. 30, no. 8, pp. 763–781, 2006.

[LAL 09] LALOUI L., NUTH M., "Investigations on the mechanical behaviour of a Heat Exchanger Pile", *5th International Symposium on Deep Foundations on Bored and Auger Piles (BAP V)*, University of Ghent, Belgium, pp. 343–347, 2009.

[LOV 12] LOVERIDGE F., POWRIE W., "Pile heat exchangers: thermal behaviour and interactions", *Proceeding of the ICE – Geotechnical Engineering*, vol. 166, no. 2, pp. 178–196, 2012.

[MCC 11] MCCARTNEY J.S., ROSENBERG J.E., "Impact of heat exchange on the axial capacity of thermo-active foundations", *ASCE Geotechnical Special Publication No.211*, ASCE, pp. 488–498, 2011.

[MOR 94] MORINO K., OKA T., "Study on heat exchanged in soil by circulating water in steel pile", *Energy and Buildings*, vol. 21, no. 1, pp. 65–78, 1994.

[NAG 05] NAGANO K., KATSURA T., TAKEDA S., *et al.*, "Thermal characteristics of steel foundation piles as ground heat exchangers", *Proceeding of the 8th IEA Heat Pump Conference*, Las Vegas, NV, pp. 6–12, 2005.

[PAH 06] PAHUD D., HUBBUCH M., "Measured thermal performances of the Dock Midfield energy pile system at Zurich airport", *14 Schweizerisches Status-Seminar, Energy und Umweltforschung im Bauwesen*, ETH Zurich, pp. 217–224, 2006.

[PAH 07] PAHUD D., HUBBUCH M., "Measured thermal performances of the energy pile system of the Dock Midfield at Zurich airport", *Proceeding of the European Geothermal Congress*, Unterhaching, Germany, 2007.

[PRA 09] PRALLE N., FRANZIUS J-N., ACOSTA F., *et al.*, "Using tunneling concrete segments as geothermal energy collectors", *Proceedings of the 5th Central European Congress on Concrete Engineering*, Baden, pp. 137–141, 2009.

[SCH 06] SCHNÜRER H., SASSE C., FISCH M.N., "Thermal energy storage in office buildings foundations", *Proceedings of the 10th International Conference on Thermal Energy Storage*, ECOSTOCK, Galloway, NJ, 2006, available at http://intraweb.stockton.edu/eyos/energy_studies/content/docs/FINAL_PAPERS/11A-4.pdf.

[SCH 12] SCHNEIDER M., MOORMANN C., VERMEER P., "Experimentelle und numerische Untersuchungen zur Tunnelgeothermie", *8. Kolloquium Bauen in Boden und Fels,* TAE, Ostfildern, 2012, available at http://www.uni-stuttgart.de/igs/content/publications/219_TAE_Schneider.pdf.

[SEK 05] SEKINE K., OOKA R., YOKOI M., *et al.*, "Development of a ground source heat pump system with ground heat exchanger utilizing the cast-in-place concrete pile foundations of a building", *Proceedings of the World Sustainable Buildings Conference*, Tokyo, pp. 1059–1066, 2005.

[SEK 06] SEKINE K., OOKA R., HWANG S., *et al.*, "Development of a ground source heat pump system with ground heat exchanger utilizing the cast-in-place concrete pile foundations of buildings", *Proceedings of the 10th International Conference on Thermal Energy Storage*, ECOSTOCK, Galloway, NJ, 2006, available at http://intraweb.stockton.edu/ eyos/energy_studies/content/docs/FINAL_PAPERS/11A-3.pdf.

[STE 12] STEWART M., MCCARTNEY J., "Strain distributions in centrifuge model energy foundations", *ASCE Geotechnical Special Publication No. 225*, ASCE, pp. 4376–4385, 2012.

[WAN 12] WANG B., BOUAZZA A., BARRY-MACAULAY D., *et al.*, "Field and laboratory investigation of a heat exchanger pile", *ASCE Geotechnical Special Publication No.* 225, ASCE, pp. 4396–4405, 2012.

[WOO 09] WOOD C.J., LIU H., RIFFAT S.B., "Use of energy piles in a residential building, and effects on ground temperature and heat pump efficiency", *Géotechnique*, vol. 59, no. 3, pp. 287–290, 2009.

[WOO 10] WOOD C.J., LIU H., RIFFAT S.B., "An investigation of the heat pump performance and ground temperature of a piled foundation heat exchanger system for a residential building", *Energy*, vol. 38, no. 12, pp. 4932–4940, 2010.

[XIA 12] XIA C., SUN M., ZHANG G., *et al.*, "Experimental study on geothermal heat exchangers buried in diaphragm walls", *Energy and Buildings*, vol. 52, pp. 50–55, 2012.

References	Location	Ground profile				Pile details			
		Depth (m)	Description	GWL (m)	Temperature (°C)	Type	Diameter (mm)	Length and spacing (m)	Number
[HAM 07]	Sapporo, Japan		Unknown		9	Hollow pre-cast concrete; mortar filled	300	9.0 and 1.2–5.7	26
[HEN 98]	Geneva, NY, USA	0–6.7	Fine silty sand with clay	near 0	8	Steel tube with concrete infill	200	25.9 and 0.6–4.6	198
		6.7–26	Glacial till with boulders						
		26–27	Fine to course sand						
		>27	Glacial till with boulders						
[KIP 09] [SCH 06]	Berlin, Germany	?	Made ground	near 0	10			8.5 and unknown	196
		?	Soft/loose lacustrine soil						
		?	Glacial sand and gravel						
[KIP 09]	Unknown, Germany		Unknown			Unknown		20 and unknown	101
[PAH 06] [PAH 07]	Zürich Airport, Switzerland	0–6	Sand and gravel	near 0	12	Bored, cast *in situ* concrete	900–1,500	30 and 9.0 (avg.)	306
		6–30	Soft lacustrine silt						
		>30	Glacial moraine						
[KIP 09] [SCH 06]	Berlin, Germany	?	Made ground	?	10	Raft foundation	8,000 m^2 absorber surface		
		?	Glacial sand andgravel						

Table 3.4a. *Summary of operational system observations – ground conditions and geostructure details*

References	Location	Thermal characteristics						
		No. loops per pile	Duration (days)	Input temperature (°C)	Flow rate (l/hr)	Power output (W/m pile)	Conductivity (W/m.K)	COP/SCOP
[HAM 07]	Sapporo, Japan	1	130 days	17.8	263	54	–	–
		2		18.9	244	55	–	–
		1		–0.6	-	69	–	3.9/3.2
[HEN 98]	Geneva, NY, USA	1	2 years	3–30	-	18.3 injection 16.4 extraction	–	3.9–4.1
[KIP 09] [SCH 06]	Berlin, Germany		4 years			35 injection 30 extraction	–	5.8
[KIP 09]	Unknown, Germany		4 years			35 injection 30 extraction	–	3.5–5.0
[PAH 06] [PAH 07]	Zürich Airport, Switzerland	5	2 years	falls to c. 3	860	45 (72 pk) extraction	–	5.1
				const. at 17		16 (33 pk) injection	–	
[KIP 09] [SCH 06]	Berlin, Germany	21.3	4 years	–	–	5 W/m^2	–	2.5–5.0

Table 3.4b. *Summary of operational system observations – thermal performance*

References	Location	Ground profile				Pile details			
		Depth (m)	Description	GWL (m)	Temperature (°C)	Type	Diameter (mm)	Length and spacing (m)	Number
[BRE 10]	Richmond, TX, USA	0–10	Stiff to very stiff clay	3.3	22	Auger pressure grouted piles	1× 300 1× 450	18.3 and 4.5	3
		10–17	Very dense sand						
		17–18.5	Stiff to very stiff clay						
		>18.5	Medium to very dense sand						
[JAL 11]	Saga City, Japan	0–15	Soft clay; m/c 12–173%	near 0	17	Steel screw with sand/water infill	140	20 and 10–18	3
		15–20	Sand and sandy Clay; m/c 30–150%						
[MOR 94]	Funabashi, Japan	0–4.5	Clayey fine sand	1.1	15	Hollow steel tube; water infill	400	20	1
		4.5–10.5	Silty clay/clayey silt						
		10.5–17	Fine sand						
		>17	Silty sand						
[NAG 05]	Hanamaki, Japan	0–8	Clayey fill with gravel	>8	~13	Hollow steel tube; water infill?	145	8.0 and 1.8	25
		>8	Clayey fill with gravel						
[SEK 05] [SEK 06]	Chiba, Japan	0–4.9	Loam/clayey loam	11	15	Bored, cast *in situ* concrete	1,500	20 and 6.0	2
		4.9–5.6	Clay (Tuff)						
		5.6–7.7	Clay/silty clay						
		>7.7	Fine sand						
[WOO 09] [WOO 10]	Burton-on-Trent, UK	0–3	Made ground (granular)	c. 3	11	Augered, cast *in situ* concrete	300	10 and 1.85–2.45	16
		>3	Soft clay						

Table 3.5a. *Summary of field test observations with ground temperature field – ground conditions and geostructure details*

Reference	Location	Thermal characteristics						
		No. loops per pile	Duration (days)	Input temperature (°C)	Flow rate (l/hr)	Power output (W/m pile)	Conductivity (W/m.K)	COP/SCOP
[BRE 10]	Richmond, TX, USA	2	3–4	40–45	–	–	2.9–3.3	–
[JAL 11]	Saga, Japan	1 (sand)	1	27	120, 240, 480	25–32	–	–
		1 (water)	1			37–55	–	–
		4? (sand)	1			27–40	–	–
[MOR 94]	Funabashi, Japan	1	3x 9 h	30–40	3,900	120–140 injection	–	–
			3x 4 h	10–3	1,800	70–85 extraction	–	–
[NAG 05]	Hanamaki, Japan	1	50	0	360	20	–	–
[SEK 05] [SEK 06]	Chiba, Japan	8	68		1,980	2003: 110 injection 2004: 210 injection	–	2–6 5.5–3.7
			100			2003: 50 extraction	–	1.5–4.5
[WOO 09] [WOO 10]	Burton-on-Trent, UK	1	200	–	–	26	–	3.5

Table 3.5b. *Summary of field test observations with ground temperature field–thermal performance*

Reference	Location	Ground profile				Pile details			
		Depth (m)	Description	GWL (m)	Temperature (°C)	Type	Diameter (mm)	Length and spacing (m)	Number
[GAO 08a]	Shanghai, China	20–27	Loose sandy silt and soft silty clay	c. 1.5	18	Bored, cast *in situ* concrete	600	25 and >15 m	4
		>20–27	Firm to stiff clay						
[KAT 09]	Ishikari, Japan	0–8	Fine sand with silt layers	1–2	9	Hollow pre-cast concrete, water infill	500	25 and 6.0	1
		8–15	Sand and gravel interbedded with layers of fine sand/silt			Hollow steel tube, water infill	267	25 and6.0	3
							400		5
		>15	Silt				600, 800, 1,200	15	1
[NAG 05]	Sapporo, Japan	0–3.5	"Surface soil" (fill)	2.6	9	Hollow steel tube; water/ glycol infill	165	40 and 10	2
		3.5–7.7	Clay/peat						
		7.7–17	Fine sand				400		2
		17–31	Silt (weathered ash/lahar deposits)						
		>31	Gravel (57–94 m/y flow)						
[XIA 12]	Shanghai, China	20–27	Loose sandy silt and soft silty clay	c. 1.5	18	Diaphragm wall	1,000 thk.	38	-
		>20–27	Firm to stiff clay						

Table 3.6a. *Summary of field test observations with pile/circuit temperature only – ground conditions and geostructure details*

References	Location	THERMAL CHARACTERISTICS						
		No. loops per pile	Duration (days)	Input temperature (°C)	Flow rate (l/hr)	Power output (W/m pile)	Conductivity (W/m.K)	COP/SCOP
[GAO 08a]	Shanghai, China	2 (W)	–	32, 35, 38	342, 684, 1,026	70–94	5.8–6.2[1]	–
		1	–			58	3.9	–
		2	–			90	5.8	–
		3	–			108	6.9	–
[KAT 09]	Ishikari, Japan	1	91	2/25	480, 960	70–90	2.6–3.0	–
		1			480, 960, 1,440		3.0–3.7	–
		1	13x 12	25–30			4.5–4.8	–
[NAG 05]	Sapporo, Japan	1	18–62	2	600–1,800	15–20	2.0–2.7	
		1					2.4–3.5	
[XIA 12]	Shanghai, China	–	2 per test	32 35 38	700	34–43[2] 50–74[3] 68–99[4]	–	–

[1] Quoted as "Heat transfer coefficient".
[2] Double U-tube with 150 mm tube spacing, excavation and soil sides.
[3] Double U-tube with 150 mm tube spacing excavation side and 750 mm spacing on soil side.
[4] Single U-tube with 750 mm spacing soil side only.

Table 3.6b. *Summary of field test observations with pile/circuit temperature only–thermal performance*

Reference	Location	Ground profile				Pile details			
		Depth (m)	Description	GWL (m)	Temperature (°C)	Type	Diameter (mm)	Length and spacing (m)	Number
[AMI 08] [BOU 09]	London, UK	0–1.5	Made ground (granular)	3.0	18	Bored, cast-in-place concrete	600	23	1
		1.5–4.0	Sand and gravel						
		>4.0	Stiff, silty clay					30	1
[LAL 03] [LAL 06]	Lausanne, Switzerland	0–5.5	Very soft alluvial clay	0.0	10	Bored, cast-in-place concrete	900	25.8	1
		5.5–12	Very soft alluvial clay						
		12–21.7	Loose sandy gravelly moraine						
		21.7–25.3	Stiff "Bottom moraine"						
		>25.3	Sandstone						
[BRA 06]	Badschallerbach, Austria	>2 - 5	Colluvium (silty sand/clayey silt)	4–5	10	Bored, cast-in-place concrete bearing andwall piles	1,200	9 and 2–8	143
		<10–15	Weathered, heavily jointed clayey sandy silt						
		>10–15	Clayey sandy silt						
[BRA 06]	Unknown	Unknown	Weak silt	near 0	–	Driven, pre-cast concrete	0.4 x0.4	24 and 1.4–4.0	570
		Unknown	Sandy gravel						

Table 3.7a. *Summary of field test with thermomechanical observations – ground conditions and geostructure details*

References	Location	Thermal characteristics						
		No. loops per pile	Duration (days)	Input temperature (°C)	Flow rate (l/hr)	Power output (W/m pile)	Conductivity (W/m.K)	COP/SCOP
[AMI 08]	London, UK	2	42	–6.0	–	–	1.52	–
[BOU 09]		4	42	+56		–		
[LAL 03] [LAL 06]	Lausanne, Switzerland	4	30	15–20	–	–	–	–
[BRA 06]	Badschallerbach, Austria	–	–	–	–	–	–	–
[BRA 06]	Unknown	–	–	–	–	65	–	–

Table 3.7b. *Summary of field test with thermomechanical observations– thermal performance*

Chapter 4

Behavior of Heat-Exchanger Piles from Physical Modeling

4.1. Introduction

Heat-exchanger piles are one of the sustainable technologies for intermittent energy storage in soils. They consist of a pile foundation equipped with a tube or a pipe network through which a fluid flows in order to exchange heat with the surrounding soil. Although this technology has been widely used recently in Austria, Germany, Switzerland and the United Kingdom, it is still rarely used in other countries, mainly due to the absence of reliable technical assessment and guarantees. Actually, the design of geothermal piles is derived primarily from the building energy demands and the thermal properties of its main components. There is currently a lack of established calculation method for the geotechnical design of geothermal piles. Improved knowledge on geotechnical design methods and the mechanisms induced by heat transfer is at the forefront of research today in geotechnical engineering. Recent studies include *in situ* tests, numerical simulations and physical modeling.

In this chapter, first, the physical modeling of pile foundation is analyzed. Discussions focus on boundary conditions and monitoring techniques in order to represent the field conditions and to capture the mechanical behavior of the foundation. Second, the physical modeling of a heat-exchanger pile is presented.

Chapter written by Anh Minh TANG, Jean-Michel PEREIRA, Ghazi HASSEN and Neda YAVARI.

The main observation on the mechanical behavior under thermomechanical loading is analyzed. In addition, heat transfer monitored during experiments is also presented. In addition to the research based on *in situ* testing and numerical simulation, these results obtained from physical modeling give complementary understanding of the behavior of heat-exchanger piles under thermomechanical loading.

4.2. Physical modeling of pile foundations

The mechanical behavior of pile foundation has been largely studied using various approaches: physical modeling, *in situ* testing, numerical modeling, etc. Among these methods, physical modeling has been considered from the beginning. As an example, Cooke and Whitaker [COO 61] performed model experiments in a laboratory where a pile of a few centimeters in diameter was tested. The pile was embedded in a soil cylinder that represents the surrounding soil in the field. To study the mechanical behavior of the pile, axial load was applied to the pile head and the penetration of the pile inside the soil was measured. The principle of such physical modeling is simple and this method can give a general view of the pile foundation behavior. After Whitaker [WHI 57], the testing of small-scale models is an accepted practice in other branches of engineering. In the following sections, a selection of existing experiments is presented in detail. Application of these techniques for testing heat-exchanger piles is then discussed.

4.2.1. *Boundary conditions*

In several cases, soil is contained in a tank with rigid walls [WHI 57, EL 10]. The deformation of the tank's walls is considered negligible under mechanical loading of the pile. The soil surface is free of stress. One of the drawbacks of these methods is that the stress state in soil in the experiments is significantly lower than that in the field. This can affect the soil–pile interaction since the mechanical behavior of the soil and soil–pile interface can be influenced by the stress range. To overcome this difficulty, in the experiments by Ergun and Akbulut [ERG 95], a surcharge air pressure of 50 kPa was applied to the soil surface via a rubber membrane and kept constant during testing. This system is similar to that used by Yasufuku and Hyde [YAS 95]. Such a calibration chamber system can then induce high stress states during the laboratory experiments, leading to conditions closer to those in the field. Calibration chambers have been widely used in foundation studies. The possibility of doing tests under known stress–strain histories and also controlled boundary conditions makes them interesting tools in simulating *in situ* tests [HOL 91]. However, because of the limited size of the chamber, the measured bearing capacities may be different from the real bearing capacities. As a result, the diameters of the pile and the chamber should be chosen in a way that minimizes boundary effects [PAI 04]. According to Parkin and Lunne [PAR 82], in order to minimize size effects, the ratio between chamber diameter and pile diameter should be at least 20 in loose sands and 50 in dense sands. Calibration chambers could be of either rigid or flexible wall type [CHI 96, WEI 08]. It is not easy to control lateral stress in rigid-walled chambers, so very large chambers are required to minimize the effects of the wall type [WEI 08]. According to Holden [HOL 91], better simulation of stress and strain is possible in smaller but flexible chambers. The relationship between the pile dimensions and the grain size (known as scale effects) could also influence the representativeness of the model. Studies show that lateral friction is independent of this effect when the pile diameter is 100 times greater than the median value of the grain size [WEI 08]. Even if calibration chambers are usually used in foundation studies, this method is still not used to study energy piles.

The centrifuge could be used in small-scale modeling of geotechnical structures (such as deep foundations) for which gravity is a primary driving force. According to Sakr and El Naggar [SAK 03], the main advantage of centrifuge modeling is that this method can simulate the linear increase in the effective stress with depth. Centrifuge modeling has been widely used in investigating pile performance in different soils [HOR 03, FIO 11]. To study the response of thermoactive foundations, this method has been used by McCartney and Rosenberg [MCC 11, Chapter 5].

For testing heat-exchanger piles, the boundary conditions related to the heat transfer should also be considered. We could choose a soil container large enough to consider that the temperature at the tank's wall and bottom does not induce boundary effects on the observed thermomechanical behavior of the pile. As another precaution, the tank's wall and bottom must be thermally insulated. The heat transfer at the soil surface should also be considered.

4.2.2. *Mechanical loading system*

Whitaker [WHI 57] used a dead load to apply an axial load on the pile head. This system was replaced by a hydraulic jack in the work of El Sawwaf [EL 10], or by an actuator as in the works of Yasufuku and Hyde [YAS 95] and Fioravante [FIO 11]. Kalantidou *et al.* [KAL 12], in their small-scale study on geothermal piles, applied dead weights on the pile and conducted thermal loading phases within constant load at the pile head; this condition is closer to that of energy piles. The use of a hydraulic jack or mechanical press would require more complex control.

4.2.3. *Monitoring*

Besides the axial load applied to the pile head, various methods have been applied to measure the stress along the pile during experiments. Choy *et al.*

[CHO 07] used a model pile composed of a series of segments equipped with cylindrical load cells to measure the axial load distribution. This technique will not be suitable for testing energy piles because of the difficulty related to the installation of the temperature control system along the pile. The second method corresponds to a continuous instrumented model pile made from a single tube. Strain gauges are fixed externally along the pile to measure axial strain at various locations along its length. All strain gauges are assembled in full Wheatstone bridges to produce a temperature-compensated system. This technique has been used by Rosquoet [ROS 04] and Choy *et al.* [CHO 07]. In the works of Horikoshi *et al.* [HOR 03] and Amatya *et al.* [AMA 06], the strain gauges are fixed internally along the pile. However, the fabrication of such a model pile becomes more complex.

For studying the mechanical behavior of energy piles, comparing to other techniques such as installing load cells in the pile [FIO 11] or measuring the friction between pile and soil [WEI 08], the full-bridge strain gauges have the following two advantages: (1) the geometry of the pile, a single metallic tube, is not modified so that the behavior of the pile during thermal phases is not influenced by the installation of the sensors; (2) the thermal dilation/contraction of the pile being isotropic, the full-bridge configuration of the strain gauges allows the effects of temperature changes on the measurement of axial strain to be minimized.

Besides the measurement of stress state along the pile, the behavior of soil around the pile can also be observed. Ni *et al.* [NI 10] used transparent soil and particle image velocimetry to study pile penetration in clay in a physical model. This allows obtaining the displacement field of soil around the pile. In the works of Jardine *et al.* [JAR 09] and Zhu *et al.* [ZHU 09], the normal stress was measured at various positions in the soil. Nevertheless, several works showed that the measurement of normal stress in soil using total pressure sensors is not always straightforward [JAR 09], especially when a 1 g physical model is considered and the stresses measured are very low (not more than a few kilopascals). Actually, when using calibration chambers or centrifuges, the stress in physical model is much higher [MCC 11, FIO 11, PAI 04].

4.2.4. *Pile's behavior*

As far as the pile's behavior is concerned, the main features usually observed for bored piles under static axial loading are (1) progressive mobilization of the friction on the pile's surface [ZHA 98, BON 95, KON 87, MOH 63]; (2) increase in axial force at all levels during loading [MOH 63]; and (3) increase in creep rate with axial load [BON 95, KON 87, MCC 70]. In the case of energy piles, McCartney and Rosenberg [MCC 11] performed static axial loading test on pile at various temperatures. The results allow us to determine the effect of temperature on the load-settlement curve. Nevertheless, it would also be interesting to test the mechanical behavior of the pile under thermal loading at various stress states. This condition is closer to the real conditions of energy piles where the pile will be heated and cooled under the load of the building.

4.3. Physical modeling of a heat-exchanger pile

4.3.1. *Experimental setup*

A physical model has been developed within the project PiNRJ (funded by the French Research Agency) to study the behavior of a model energy pile. The experimental setup is shown in Figure 4.1 and the details on dimensions and sensor installation are shown in Figure 4.2. For loading the pile on its head, a progressive loading is adopted by flowing water into the water tank that is placed at the upper part of the structure. By letting more water enter the tank, the applied load on the pile head will increase. The force sensor installed at the pile head measures the axial load applied to the pile head. A displacement transducer measures the pile head displacement.

The model pile is a closed-end aluminum cylinder with outer and inner diameters of 20 and 18 mm, respectively. Five full-bridge strain gauges (G1–G5, Figure 4.2) are stuck to the outer surface of the pile at 100, 200, 300, 400 and 500 mm from the pile toe, respectively. Three temperature transducers (T1, T2 and T3) are also attached along the pile at 100, 300 and 500 mm from the pile toe, respectively. To have the maximum roughness between the pile and the surrounding soil, a layer of sand was coated on the outer surface of the pile.

The dimensions of the soil container are 548 mm in diameter and 880 mm in height. Considering the dimensions of the pile, the ratio of diameter of soil specimen to the pile diameter is higher than 20. Thus, the size effect is negligible according to Parkin and Lunne [PAR 82]. The distance between the pile toe and the base of the container is 250 mm, which is 12.5 times greater than the pile diameter.

To control the temperature of the pile, a temperature-controlled circulating bath is used for both heating and cooling purposes. The device allows a temperature control from –20°C to 150°C. The internal reservoir is filled with a fluid (water in this case) during thermal cycles. This circulating bath is connected to the U-shaped tube that was inserted inside the pile (Figure 4.1), the latter being filled with water to improve the homogeneity of temperature inside the pile during the tests. The U-shaped tube has an internal diameter of 2 mm.

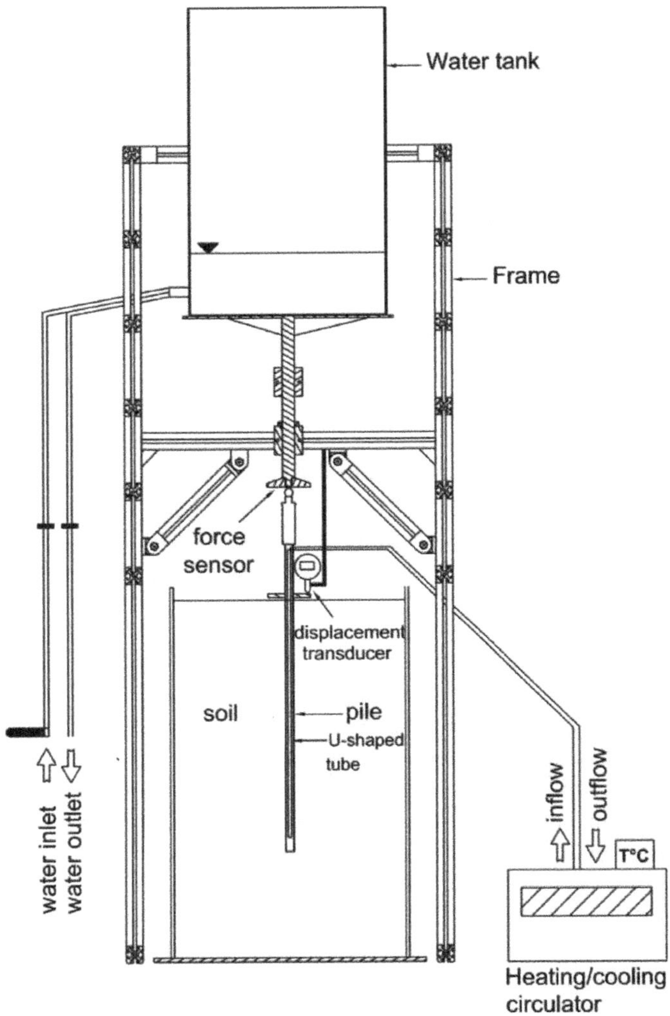

Figure 4.1. *Physical model of heat-exchanger pile: schematic view*

To monitor the temperature, besides the three temperature transducers that were stuck on the surface of the pile (T1, T2 and T3), one temperature transducer was placed inside the pile (S1), at 300 mm from the pile toe (Figure 4.2). The other temperature transducers (S2–S12) were distributed at different levels in the soil. To measure the normal stress in soil during the tests, 10 pressure transducers were installed at different positions (Figure 4.2). These sensors, which have a diameter of 6 mm and a thickness of 0.8 mm, can measure the normal stress applied on their sensing surface in the range of 0–100 kPa.

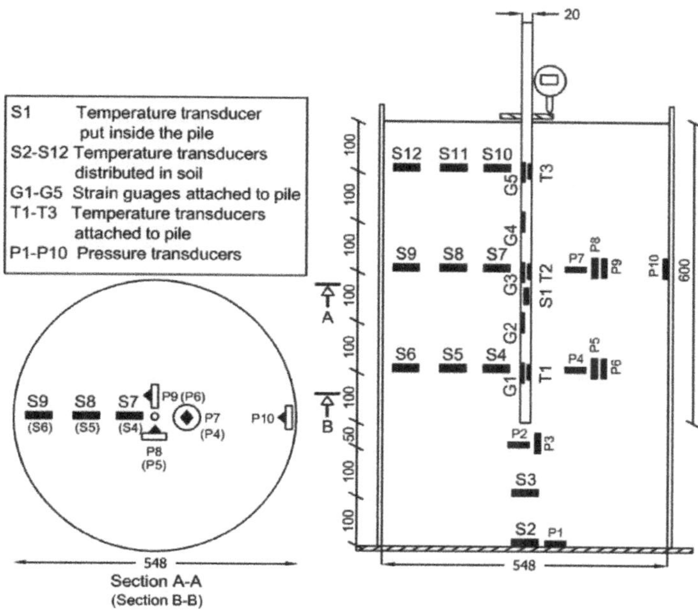

Figure 4.2. *Physical model of heat-exchanger pile: instrumentation (lengths given in millimeters)*

4.3.2. *Mechanical behavior of a pile under thermomechanical loading*

Kalantidou *et al.* [KAL 12] used a similar experimental setup to study the mechanical behavior of the pile during heating/cooling cycles. The soil used is Fontainebleau sand with a dry density of 1.51 Mg/m^3 (corresponding to a relative density of 46%). The scale effect mentioned earlier is more influenced by the ratio between the pile diameter (20 mm) and the median diameter of sand particles (D_{50} = 0.23 mm); this ratio is equal to 87. Foray *et al.* [FOR 98] propose the lower bound of 200, while Garnier and Konig [GAR 98] reduce the ratio to 100 and Fioravante [FIO 02] suggests 50. The load-settlement curve obtained allows us to estimate the conventional ultimate bearing capacity of the pile at 525 N; this load corresponds to a settlement of 10% of the pile's diameter.

Heating/cooling cycles were performed under various pile head axial loads. Figure 4.3 shows the results obtained under a constant load of 200 N (38% ultimate bearing capacity). It can be seen that incremental heating induced incremental heaves of the pile head (Figure 4.3(b)); the temperature corresponding to that measured by sensor S1 is shown in Figure 4.3(a). In Figure 4.3(c), the pile head settlement is plotted versus the pile's temperature for the two cycles. In this figure, the free thermal expansion/contraction curve of the aluminum pile is also plotted.

This reference curve corresponds to the head displacement of a pile subjected to temperature changes, when its toe is fixed. The experimental results during the first heating show a settlement/temperature slope similar to that of the pile's thermal expansion curve and subsequently the first cooling path joins the thermal contraction curve progressively. At the same time, a hysteresis phenomenon, that is distinct heating and cooling paths, can be observed in both cycles.

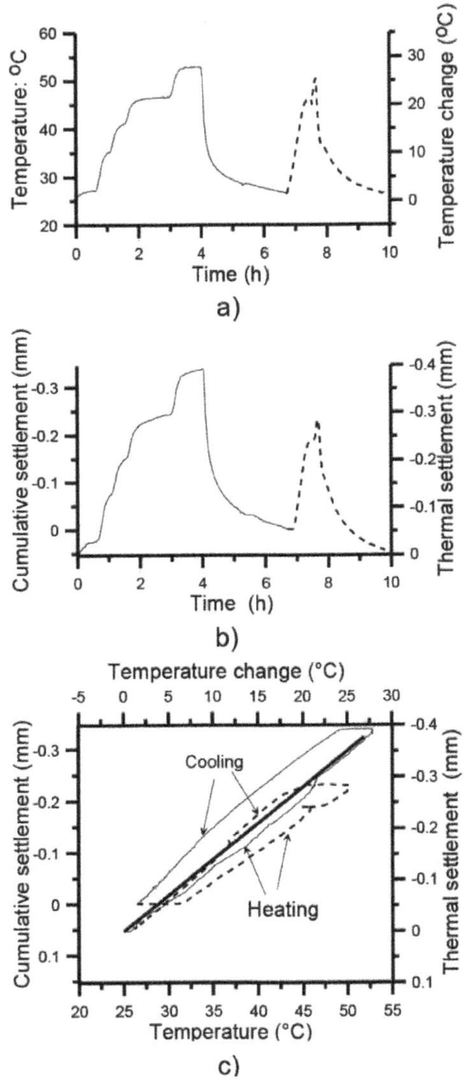

Figure 4.3. *Heating/cooling cycles under 38% ultimate bearing capacity: temperature and pile head settlement*

Figure 4.4 shows the results obtained during heating/cooling cycles at a constant load of 400 N (76% ultimate bearing capacity). The results in terms of pile head displacement show a settlement of 0.4 mm after the first cycle (Figure 4.4(b)). This phenomenon was not observed for the tests having a lower pile head load (Figure 4.3).

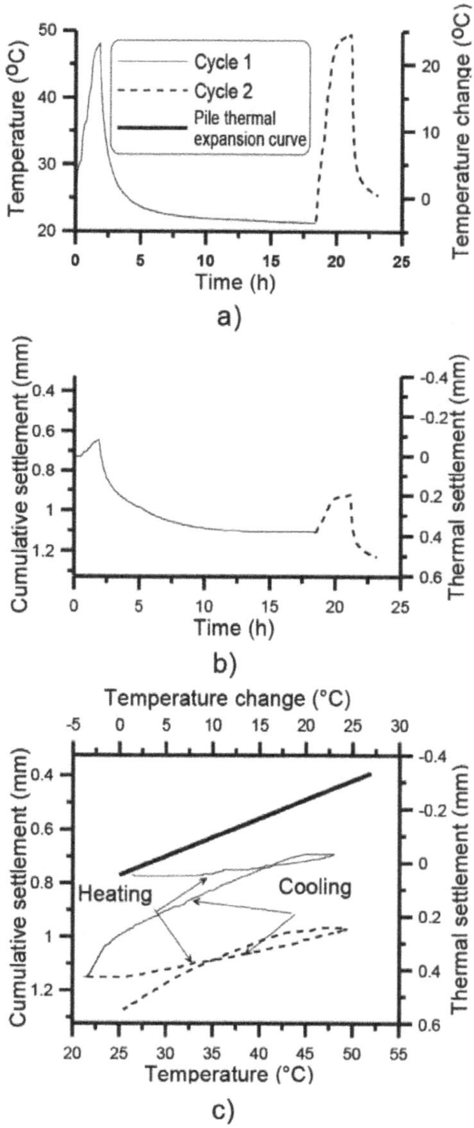

Figure 4.4. *Heating/cooling cycles under 76% ultimate bearing capacity: temperature and pile head settlement*

In Figure 4.4(c), it can be seen that the pile head heave obtained during the first heating path is significantly lower than the pile thermal expansion curve. Yet, the cooling path follows the same slope as that of the pile thermal contraction curve. The second heating induced heave and resulted in a settlement/temperature change slope relatively higher than that of the first heating path but still lower than that of the pile thermal expansion curve. On the other hand, the pile behavior during the second cooling path is similar to the first cooling path and the total settlement of the pile head after two thermal cycles is 0.5 mm.

In addition to the results obtained for the pile head settlement under thermal cycles by Kalantidou *et al.* [KAL 12], the experimental setup shown in Figures 4.1 and 4.2 allows us to monitor the axial load along the pile and other parameters in the soil. The details of one heating/cooling test performed under a constant axial load of 200 N are presented below.

In Figure 4.5, the axial forces profile obtained at various steps is shown. After applying an axial load of 200 N at the pile head, the results show that the axial force along the pile decreased progressively from 200 N at the pile head to 150 N at 500 mm depth. This corresponds to the shaft friction mobilized along the pile during mechanical loading. Heating/cooling cycles generally increased the axial forces along the pile even if the pile head load was kept constant at 200 N.

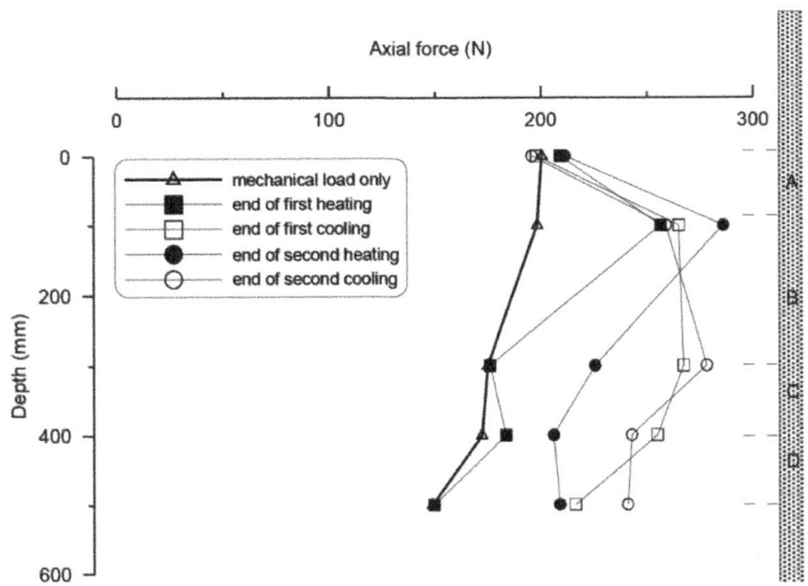

Figure 4.5. *Axial load profiles during heating/cooling cycles under an axial load of 200 N*

Figure 4.6 shows the changes of total pressures measured at various levels in the soil during heating/cooling cycles under a constant axial load of 200 N. The results show that the total pressure in soil surrounding the pile slightly changes during thermal cycles. Since the pressure range remains small (several kilopascals), the pressure change also remains small but can be captured by the sensors used.

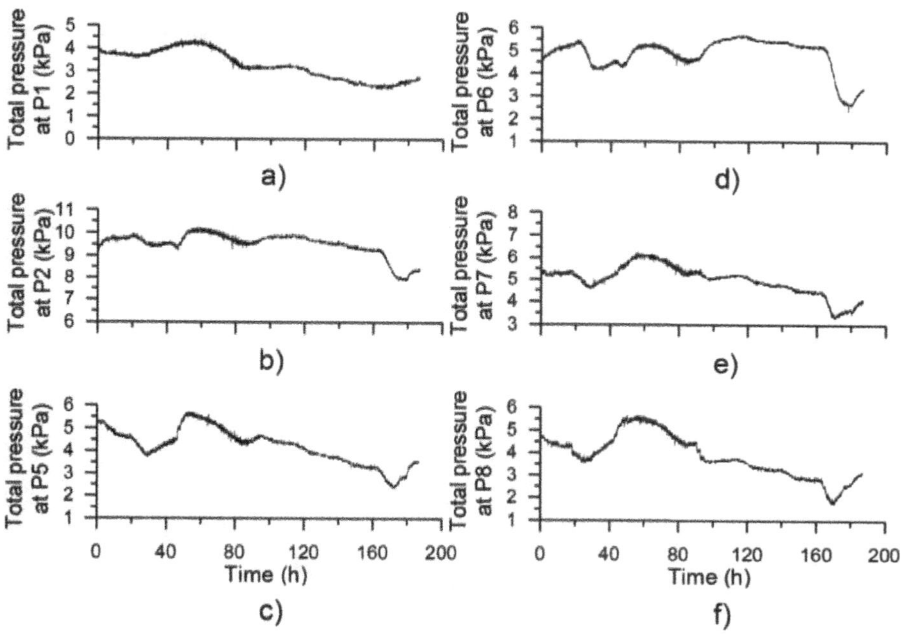

Figure 4.6. *Time evolution of total pressure during heating/cooling cycles under an axial load of 200 N*

4.3.3. *Heat transfer*

The heat transfer can be observed from the measurement of temperature at various locations along the pile and in the soil. Figure 4.7 shows the temperatures measured during a heating/cooling test. First, the temperatures measured at various locations along the pile (T1, T2 and T3) are similar to each other but slightly different from that measured inside the pile (S1) (Figure 4.7(a)). This observation evidences the heat transfer inside the model pile, from the heating/cooling tube to the pile wall. The measurements of temperature below the model pile (S2 and S3 in Figure 4.7(b)) show that the temperature changes inside the pile (between 10°C and 30°C) induced only small changes of temperature at these points (between 18°C and

20°C). For the temperature sensors around the pile (S4–S12, Figures 4.7(c)–(e)), the smaller the distance to the pile is, the higher the temperature variation. These measurements confirm the radial diffusion of heat in the soil during thermal cycles.

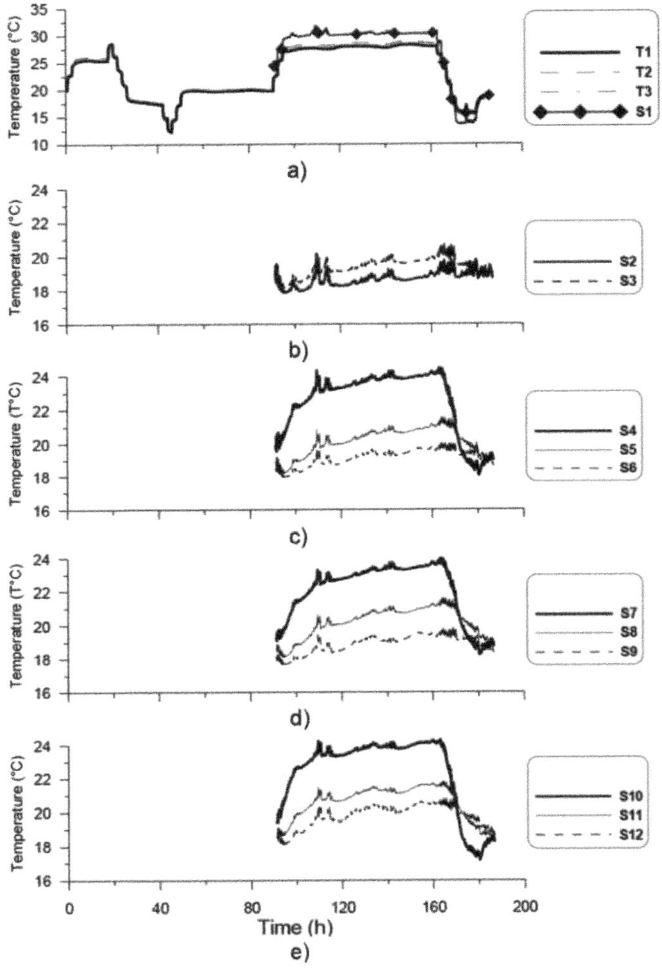

Figure 4.7. *Temperature changes during heating/cooling cycles under an axial load of 200 N*

4.3.4. *Soil–pile interface*

The results obtained on the axial load profiles allow us to determine the mobilized friction at various zones (labeled as A–D in Figure 4.5) along the pile during mechanical and thermal phases. In Figures 4.8 and 4.9, the mobilized friction at various zones is plotted against the pile head settlement. The results show that friction was mobilized all along the pile's length during the mechanical loading. The mobilized friction seems to be higher at larger depths. Heating/cooling cycles changed the friction mobilization significantly and the direction of the mobilized shaft friction at zones A and B was inverted. It has to be noted that for obtaining thermal displacements, mechanical displacements observed just before starting the application of thermal cycles were subtracted from the total thermomechanical displacements.

4.3.5. *Lessons learned from physical modeling of a heat-exchanger pile*

In this physical model, the pile can be assumed as undeformable under mechanical loading in the considered mechanical loading range (0–500 N). Actually, for a given load of 525 N, the pile's length reduction is equal to 0.076 mm. As a result, the measured pile head displacement during the loading phase represents the pile toe displacement. In the case of tests at lower axial loads, heating induced a thermal dilation of the pile and pile head displacement is similar to the reference thermal expansion curve (head displacement of a pile subjected to temperature changes when its toe is fixed). This means that the mobilized pile toe capacity remains small, much lower than the ultimate resistance. The axial dilation of the pile during heating would invert the direction of the mobilized shaft friction at some points (Figure 4.8). Taking into account that the soil has a tendency to restrain the dilation of the pile, additional stress can be developed at the pile toe (as explained by Laloui *et al.* [LAL 03] and Bourne-Webb [BOU 09]). Nevertheless, the additional stress induced by heating did not cause settlement of the pile toe. This is not the case for tests where higher loads were applied. Additional stress at the pile toe induced by heating led to additional settlement and the development of irreversible strains. This explains why the pile head heave obtained during heating is significantly lower than the pile thermal expansion curve (Figure 4.4) and why irreversible settlement was observed after the thermal cycles of these tests.

The results on the axial force profiles (Figure 4.5) and the mobilized friction on the pile's surface (Figure 4.8) obtained during mechanical loading show the performance of strain gauges capturing the mechanical behavior of the pile during mechanical and thermal loading stages. Note that the few existing works concern *in situ* tests using more complex techniques [LAL 06, BOU 09]. In the works of Kalantidou *et al.* [KAL 12] and [WAN 11], an energy pile was tested in a physical model but the axial force profile was not measured. The results on *in situ* tests show that these measurements are crucial in order to interpret the behavior of the pile under thermomechanical loading.

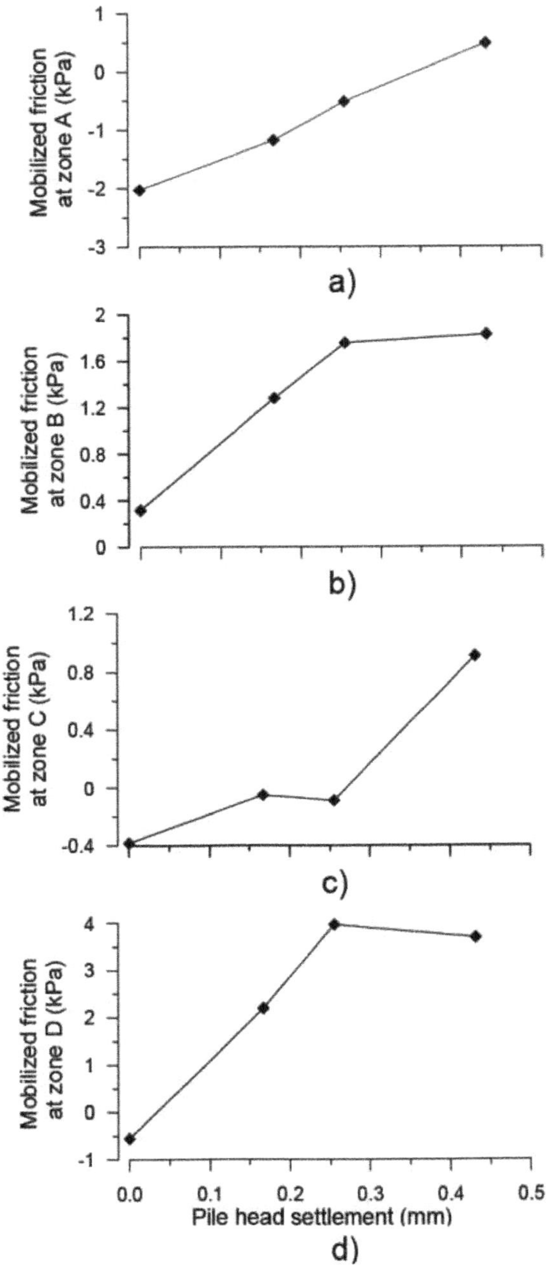

Figure 4.8. *Mobilized friction along the pile during mechanical loading*

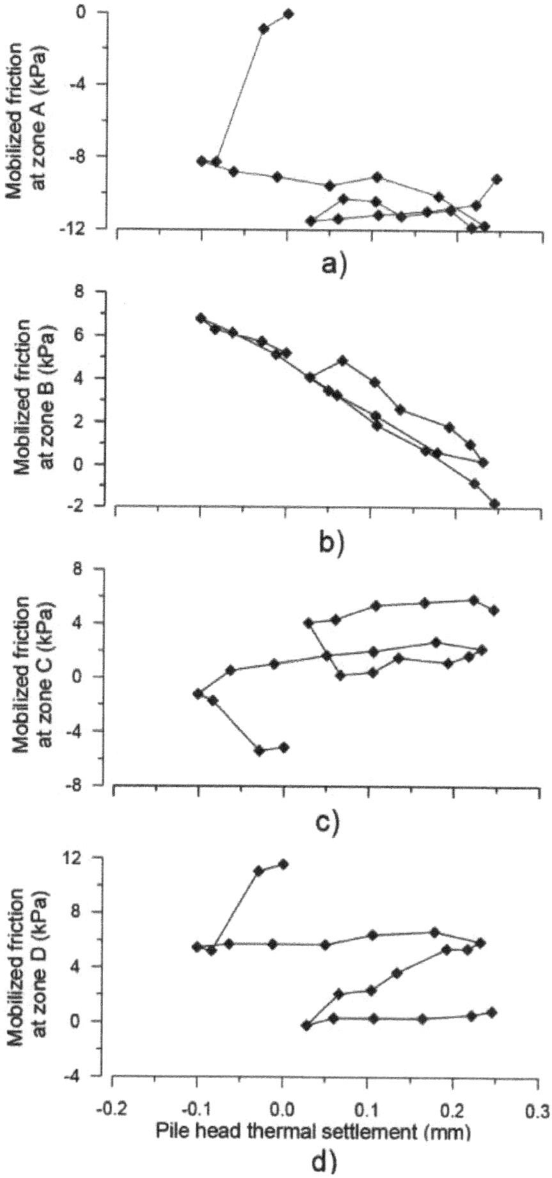

Figure 4.9. *Mobilized friction along the pile during thermal loading*

4.4. Conclusions

Physical modeling has been widely used to study the mechanical behavior of pile foundations. In section 4.2, some existing works were reviewed and the following points were discussed: boundary conditions, mechanical loading system, monitoring and pile's behavior. These points were analyzed in the perspective of using a physical model to study the mechanical behavior of the energy pile.

Section 4.3 presents an experimental study investigating the mechanical behavior of an energy pile using a physical model. The experimental setup is composed of a model pile (with a diameter of 20 mm and length of 800 mm), a soil container, a heating/cooling circulator and a loading device, which is a water tank. A number of transducers were implemented into the soil (temperature and total pressure transducers), others were mounted on the pile's surface (strain and temperature gauges). A displacement transducer and a force sensor were also installed on the pile head.

The behavior of piles subjected to thermal cycles under different constant axial loads was investigated. The pile response, in terms of pile head settlement, appears to be "thermoelastic" under thermal cycles when the mechanical load is less than about 40% of the ultimate resistance. When the mechanical load exceeds 40% of the ultimate resistance, irreversible pile settlement develops. The axial force profiles are significantly modified during thermal cycles. The results evidenced that in some zones, the mobilized shaft friction was inverted during these cycles. The changes of total pressure in the soil around the pile were also observed. Finally, measurement of temperature at various locations in the soil confirms that heat was diffused radially from the pile to the surrounding soil.

4.5. Acknowledgments

The authors would like to express their gratitude to the French National Research Agency for funding this study, which is part of the project PiNRJ "Geotechnical aspects of foundation energy piles" – ANR 2010 JCJC 0908 01.

4.6. Bibliography

[AMA 06] AMATYA B.L., TAKEMURA J., KHAN M.R.A., *et al.*, "Centrifuge modelling on performance of natural clay barrier with piles", *International Journal of Physical Modelling in Geotechnics*, vol. 6, no. 4, pp. 1–18, 2006.

[BON 95] BOND A.J., JARDINE R.J., "Shaft capacity of displacement piles in a high OCR clay", *Géotechnique*, vol. 45, no. 1, pp. 3–23, 1995.

[BOU 09] BOURNE-WEBB P., AMATYA B., SOGA K., *et al.*, "Energy pile test at Lambeth College, London: geotechnical and thermodynamic aspects of pile response to heat cycles", *Géotechnique*, vol. 59, no. 3, pp. 237–248, 2009.

[CHI 96] CHIN J.T., POULOS H.G., "Tests on model jacked piles in calcareous sand", *Geotechnical Testing Journal*, vol. 19, no. 2, pp. 164–180, 1996.

[CHO 07] CHOY C.K., STANDING J.R., MAIR R.J., "Stability of a loaded pile adjacent to a slurry-supported trench", *Géotechnique*, vol. 57, no. 10, pp. 807–819, 2007.

[COO 61] COOKE R.W., WHITAKER T., "Experiments on model piles with enlarged bases", *Géotechnique*, vol. 11, no. 1, pp. 1–13, 1961.

[EL 10] EL SAWWAF M., "Experimental study of eccentrically loaded raft with connected and unconnected short piles", *Journal of Geotechnical and Geoenvironmental Engineering*, vol. 136, no. 10, pp.1394–1402, 2010.

[ERG 95] ERGUN M.U., AKBULUT H. "Bearing capacity of shaft-expanded driven model piles in sand", *Géotechnique*, vol. 45, no. 4, pp. 715–718, 1995.

[FIO 02] FIORAVANTE V., "On the shaft friction modelling of non displacement piles in sand", *Soils and Foundations*, vol. 42, no. 2, pp. 23–33, 2002.

[FIO 11] FIORAVANTE V., "Load transfer from a raft to a pile with an interposed layer", *Géotechnique*, vol. 61, no. 2, pp. 121–132, 2011.

[FOR 98] FORAY P., BALACHOWSKI L., RAULT G., "Scale effect in shaft friction due to the localisation of deformations", *Proceedings of the International Conference Centrifuge 98, Tokyo*, vol. 1, Balkema, Amsterdam, the Netherlands, pp. 211–216, 1998.

[GAR 98] GARNIER J., KONIG D., 1998, "Scale effects in piles and nail loading tests in sand", *Proceedings of the International Conference Centrifuge 98, Tokyo*, vol. 1, Balkema, Rotterdam, pp. 205–210, 1998.

[HOL 91] HOLDEN J.C., "History of the first six CRB calibration chambers", *Proceedings of the 1st International Symposium on Calibration Chamber Testing*, Potsdam, pp. 1–11, 1991.

[HOR 03] HORIKOSHI K., MASTUMOTO T., HASHIZUMB Y., *et al.*, "Performance of piled raft foundations subjected to static horizontal loads", *International Journal of Physical Modelling in Geotechnics*, vol. 2, no. 2, pp. 37–50, 2003.

[JAR 09] JARDINE R.J., ZHU B., FORAY P., *et al.*, "Experiment arrangement for investigation of soil stresses developed around a displacement pile", *Soils and Foundations*, vol. 49, no. 5, pp. 661–673, 2009.

[KAL 12] KALANTIDOU A., TANG A.M., PEREIRA J.M., *et al.*, "Preliminary study on the mechanical behaviour of heat exchanger pile in physical model", *Géotechnique*, vol. 62, no. 11, pp. 1047–1051, 2012.

[KON 87] KONRAD J.-M., ROY M., "Bearing capacity of friction piles in marine clay", *Géotechnique*, vol. 37, no. 2, pp. 163–175, 1987.

[LAL 03] LALOUI L., MORENI M., VULLIET L., "Comportement d'un pieu bi-fonction, fondation et échangeur de chaleur", *Canadian Geotechnical Journal*, vol. 40, no. 2, pp. 388–402, 2003.

[LAL 06] LALOUI L., NUTH M.,VULLIET L., "Experimental and numerical investigation of the behaviour of a heat exchanger pile", *International Journal for Numerical and Analytical Methods in Geomechanics*, vol. 30, no. 8, pp. 763–781, 2006.

[MCC 70] MCCAMMON N.R., GOLDER H.Q., "Some loading tests on long pipe piles", *Géotechnique*, vol. 20, no. 2, pp. 171–184, 1970.

[MCC 11] MCCARTNEY J.S., ROSENBERG J., "Impact of heat exchange on side shear in thermo-active foundations", *Proceedings of the Geo-Frontiers 2011 Conference*, vol. 211, ASTM, Geotechnical Special Publications (GSP), pp. 488–498, 2011.

[MOH 63] MOHAN D., KUMAR V., "Load-bearing capacity of piles", *Géotechnique*, vol. 13, no. 1, pp. 76–86, 1963.

[NI 10] NI C., HIRD C., GUYMER I., "Physical modelling of pile penetration in clay using transparent soil and particle image velocimetry", *Géotechnique*, vol. 60, no. 2, pp. 121–132, 2010.

[PAI 04] PAIK K., SALGADO R., "Effect of pile installation method on pipe pile behaviour in sands", *Geotechnical Testing Journal*, vol. 27, no. 1, pp. 78–88, 2004.

[PAR 82] PARKIN A.K., LUNNE T., "Boundary effects in the laboratory calibration of a cone penetrometer for sand", *Proceedings of the 2nd European Symposium on Penetration Testing*, vol. 2, pp. 761–768, 1982.

[ROS 04] ROSQUOET F., Pieux sous charge latérale cyclique, PhD Thesis, Ecole Centrale de Nantes, France, 2004.

[SAK 03] SAKR M., EL NAGGAR M.H., "Centrifuge modelling of tapered piles in sand", *Geotechnical Testing Journal*, vol. 26, no. 1, pp. 22–35, 2003.

[WAN 11] WANG B., BOUAZZA A., HABERFIELD C., "Preliminary observations from laboratory scale model geothermal pile subjected to thermo-mechanical loading", *Proceedings of the Geo-Frontiers 2011 Conference*, vol. 211, ASTM, Geotechnical Special Publications (GSP), pp. 430–439, 2011.

[WEI 08] WEINSTEIN G.M., Long-term behavior of micropiles subject to cyclic axial loading, PhD Thesis, Polytechnic University, 2008.

[WHI 57] WHITAKER T., "Experiments with model piles in groups", *Géotechnique*, vol. 17, no.4, pp. 147–167, 1957.

[YAS 95] YASUFUKU N., HYDE A.F.L., "Pile end-bearing capacity in crushable sands", *Géotechnique*, vol. 45, no. 4, pp. 663–616, 1995.

[ZHA 98] ZHANG L., MCVAY M.C., LAI P., "Centrifuge testing of vertically loaded battered pile groups in sand", *Geotechnical Testing Journal*, vol. 21, no. 4, pp. 281–288, 1998.

[ZHU 09] ZHU B., JARDINE R.J., FORAY P., "The use of miniature soil stress measuring cells in laboratory applications involving stress reversals", *Soils and Foundations*, vol. 49, no. 5, pp. 675–688, 2009.

Chapter 5

Centrifuge Modeling of Energy Foundations

5.1. Introduction

Although incorporation of heat exchangers into deep foundation elements (energy foundations) helps reduce the installation costs of ground source heat exchange systems [BRA 98, ENN 01, BRA 06], an issue encountered is the potential for foundation movements due to thermal expansion and contraction of the foundation element or surrounding soil. Further, soil–structure interaction may restrain movement of the foundation, leading to generation of thermally induced stresses. Thermomechanical soil–structure interaction has been documented in several field-scale case histories [LAL 06, BOU 09, LAL 11, AMA 12, MCC 12]. Although thermomechanical soil–structure interaction analyses permit prediction of changes in axial stress or strain during heating and cooling operations [KNE 11, PLA 12], they require empirical data for calibration of model parameters and verification of predictions. This is especially the case when considering the behavior of energy piles in some soil deposits, such as soft clays or unsaturated soils. Centrifuge modeling is a useful approach to measure empirical parameters for soil–structure interaction analyses for energy foundations, as the properties of scale-model foundations and soil layers can be carefully controlled and different configurations can be considered for lower cost full-scale field testing. An additional benefit of centrifuge modeling is that scale-model energy foundations can be loaded to failure to characterize the effects of temperature on the load-settlement curve. The back-calculated ultimate side shear distribution and end bearing are useful parameters for soil–structure interaction analyses.

Chapter written by John S. McCartney.

Further, centrifuge tests on scale-model foundations with embedded instrumentation permit the measurement of thermally induced stresses and strains in the foundation, which can be used to validate soil–structure interaction analyses or finite element models. Areas of thermomechanical soil–structure interaction analysis being explored in the centrifuge are the impact of thermally induced radial stresses on the side shear resistance, end boundary condition (foundation toe and head) effects, and the impact of temperature on the stress–strain curves for end bearing and side shear resistance.

This chapter includes a review of the relevant scaling relationships and strategies to account for issues with scaling heat transfer in the centrifuge, the details of scale-model energy foundation development, and the procedures and typical results of tests used to define parameters for soil–structure interaction analyses. The tests discussed in this chapter focus on the behavior of semi-floating (i.e. the toe of the foundation is resting in soil) energy foundation models in unsaturated silt.

5.2. Background on thermomechanical soil–structure interaction

As an energy foundation is heated or cooled, it may expand or contract, respectively, depending on the end-restraint boundary conditions. For unconstrained conditions, the axial thermal strain can be calculated as follows:

$$\varepsilon_{T,free} = \alpha_c \Delta T \qquad [5.1]$$

where α_c is the coefficient of linear thermal expansion of reinforced concrete and ΔT is the change in temperature. Thermal strain is defined as positive for compression. Accordingly, α_c is negative as structural elements expand during heating (positive ΔT). The coefficient of thermal expansion of unreinforced concrete ranges from –9 to –14.5 με/°C depending on the aggregate mineralogy, while that of the steel reinforcements ranges from –11.9 to –13 με/°C [BOU 09, STE 12]. As these materials are relatively compatible, significant differential thermal strains are not expected in reinforced concrete. The value of $\varepsilon_{T,free}$ is an upper limit on the thermal strains due to heating or cooling. For constrained conditions, the actual thermal strains will be less than those predicted by equation [5.1]. In this case, the thermal axial stresses induced in an energy foundation can be calculated as follows:

$$\sigma_T = E(\varepsilon_T - \alpha_c \Delta T) \qquad [5.2]$$

where E is Young's modulus of reinforced concrete and ε_T is the actual thermal strain. For real energy foundations embedded in soil or rock, soil–structure interaction mechanisms will restrict the movement of the foundation during heating. Specifically, the side shear resistance, end bearing and stiffness restraint of the overlying building will lead to different distributions in thermally induced stresses and strains in an energy foundation [LAL 06, BOU 09, AMA 12, MCC 12].

5.3. Centrifuge modeling concepts

Centrifuge modeling relies on the concept of geometric similitude, which assumes that a full-scale prototype soil layer will have the same stress state as a model-scale soil layer that is N times smaller when spinning in a geotechnical centrifuge at a centripetal acceleration that is N times larger than that of the Earth's gravity [KO 88, TAY 95]. The centripetal acceleration generates an increased body force field in the scale-model. Geometric similitude can be employed to extrapolate the load-settlement behavior and thermal soil–structure interaction phenomena of scale-model energy foundations to those representative of full-scale prototype foundations in the real-world. By scaling the foundation dimensions by a factor of $1{:}N$ (model:prototype), strains in the foundation will scale by a factor of 1:1 and forces will scale by a factor of $1{:}N^2$ [KO 88, TAY 95]. One issue in modeling energy foundations is that the temperature does not depend on the increased body forces in the centrifuge. Spatial measurements of temperature in dry quartz sand surrounding a cylindrical heat source during centrifugation at different g-levels by Krishnaiah and Singh [KRI 04] confirm that centrifugation does not lead to a change in the heat flow process. However, if the dimensions associated with the spatial distribution of heat flow were scaled from model to prototype scale (assuming the same thermal conductivity in both cases), the time required for heat flow by conduction would be N^2 times faster in the centrifuge model ($1{:}N^2$). Savvidou [SAV 88] derived this scaling factor from the diffusion equation, which only includes scaling of the length. An implication of temperature scaling is that a greater volume of soil surrounding the model-scale foundation will be affected by changes in temperature. Soils change in volume with temperature, so if a greater zone of soil around the foundation is affected, then the effects of differential volume change of the foundation and soil may be emphasized. From this perspective, centrifuge modeling will provide a worst-case scenario. A solution to address the scaling issue is to calibrate numerical simulations of the tests using the data from model scale. However, if the goal of testing is to evaluate the impact of temperature on the load-settlement curve of the foundations, time should be provided to reach steady-state conditions. However, if the goal is to evaluate the impact of temperature on the axial strain distribution in the foundation, tests can be performed until strains stabilize while the foundation temperature is held constant. This amount of time depends on the soil type.

5.4. Centrifuge modeling components

5.4.1. *Centrifuge model fabrication and characterization*

Two scale-model, semi-floating energy foundations are evaluated in this chapter. The first (foundation A) is a simple foundation suitable for evaluation of the impact of temperature on the load-settlement curve of the foundation, while the second (foundation B) contains embedded instrumentation making it suitable for evaluation of the impact of temperature on the stress–strain distribution in the reinforced concrete. Both model foundations have lengths of 381 mm, but foundation A has a diameter of 76.2 mm while foundation B has a smaller diameter of 50.8 mm. Both foundations were tested at a g-level of 24, so they represent prototype drilled shafts having lengths of 9.1 m and diameters of 1.8 and 1.2 m, respectively. Schematics of the foundations are shown in Figure 5.1.

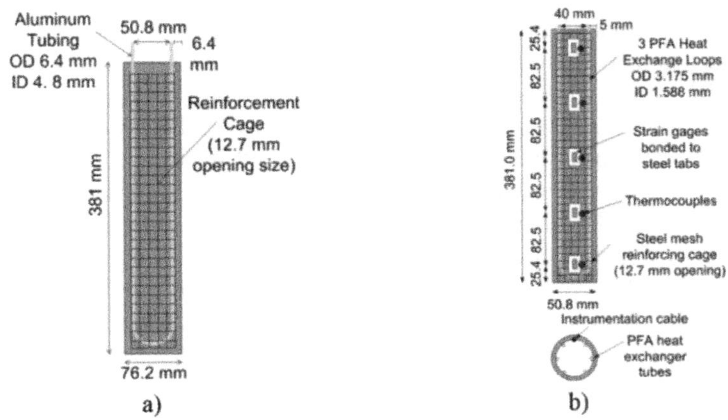

Figure 5.1. *Scale-model energy foundations: a) foundation A and b) foundation B*

Although drilled shafts are typically cast in place in the soil, the model energy foundations were precast outside the soil layer due to the large amount of instrumentation, cables and heat exchanger tubing. This also permits the foundations to be reused in subsequent tests, and to be tested outside the soil layer to characterize their thermal and mechanical properties. The foundations were formed by pouring concrete into a cardboard tube containing the reinforcement cage. The fine aggregate was uniform sand with a particle size of 0.5 mm, while the coarse aggregate was gravel having a maximum particle size less than 6 mm (to permit flow through the reinforcement cage openings). Foundation A contained a single aluminum heat exchanger tube with relatively large diameter attached inside the reinforcement cage, while foundation B contained three small-diameter, high-density perfluoroalkoxy (PFA) tubing loops attached inside the reinforcement cage. The use of three loops in foundation B permits more uniform distribution of heat. Foundation B contains five strain gages bonded to steel tabs embedded in the concrete, with thermocouples installed at the same locations. The gages were selected because their coefficient of thermal expansion was similar to that of steel, and because they have a stable response to cyclic heating and cooling. The gages were first attached with temperature resistant M-Bond AE-15 adhesive to 30 mm long steel tabs having a dog-bone shape with a hole punched at either end. This adhesive was chosen because it should be cured at an elevated temperature of 85°C, making it less likely to slip during cyclic heating and cooling than other adhesives that cure at room temperature.

A comprehensive set of characterization tests was performed on foundation B outside the soil in a load frame at 1-gravity to determine the mechanical and thermal properties of the reinforced concrete. The results from these tests are reported in detail by Stewart [STE 12]. The first test involved application of incremental axial loads under room temperature conditions, taking care to properly level the foundation and center the load to avoid bending. Young's modulus determined using the strain gauge data and a linearly variable deformation transformer (LVDT) was 7.17 GPa. This value is low compared to the Young's modulus of concrete in drilled shafts (approximately 30 GPa) due to the smaller aggregate size. The foundation was then heated to a temperature of 62°C by circulating fluid through the heat exchange tubes within the foundation. The foundation was permitted to freely expand under this axial stress, permitting definition of the coefficient of linear thermal expansion of the reinforced concrete ($\alpha_c = -7.5$ με/°C, where με is micro-strain). The thermal response of each strain gauge was different, likely due to differences in curing of the adhesive bonding the strain gages to the steel tabs. However, because the thermal strain should theoretically be the same at each location along the length of the foundation for free expansion, thermal correction factors were defined using the reading from the LVDT. Before application of any correction factors, the gages were corrected for the thermal offset error specific to this batch of gages and for differential expansion of the steel tabs (measured to be $\alpha_s = -8.5$ με/°C) and reinforced concrete [STE 12].

5.4.2. *Experimental setup*

A schematic of the containers used for testing of Foundations A and B is shown in Figures 5.2(a) and (b), respectively. In both cases, the container is a cylindrical aluminum tank with an inside diameter of 0.6 m, wall thickness of 13 mm and a height of 0.54 m. A 13 mm thick insulation sheet was wrapped around the container to prevent heat transfer through the sides of the cylinder (no-flow boundary). The bottom of the container permits some loss of heat, but it was preferred not to install insulation beneath the container to provide a stiff platform for loading. Loads were applied to the foundations using a horizontally mounted, brushed DC electric motor with a coupling to a vertical worm drive. The motor is mounted on a reaction frame. Applied loads were measured using a load cell attached to the shaft of the worm drive, and a National Instruments motor control module was used to maintain a constant load using a force-feedback control loop. Additional pictures of the container and load frame are shown in [STE 12]. The configurations of the tests were slightly different. The goal of the tests on foundation A was to measure the impact of temperature on the load-settlement curve, so four foundations were installed in the same container so that they would have nearly identical soil conditions. The spacing in Figure 5.2(a) led to minimal interference between foundations with respect to thermal loading and mechanical loading. The goal of the tests on foundation B was to characterize stress–strain distributions for validation of numerical simulations, so the foundation was placed in the middle of the container.

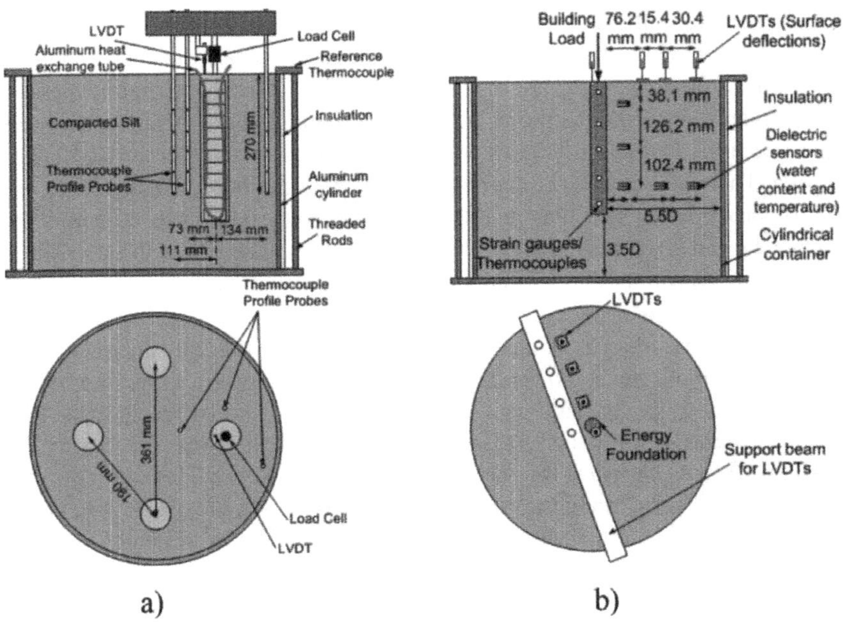

Figure 5.2. *Centrifuge-scale testing setups. a) For evaluating the capacity of energy foundations and b) for evaluating thermally induced strain distributions*

The location of instrumentation incorporated into the centrifuge containers is shown in Figure 5.2. An LVDT was placed on the top of the foundation. The LVDTs were mounted on cantilever arms connected to a support beam across the top of the container. Three thermocouple profile probes were inserted into the soil at different radial locations around foundation A. These probes were used to measure transient changes in temperature of the soil surrounding the foundation to assess heat transfer processes. Dielectric sensors (model EC-TM from Decagon Devices), capable of inferring the volumetric water content and temperature of the soil, were placed in the soil layer around foundation B during compaction.

The F25-ME refrigerated/heated circulator manufactured by Julabo, Inc., operating outside the centrifuge, was connected to foundation B via the hydraulic slip ring stack as shown in Figure 5.3. The heat pump has a working temperature range of –28°C to 200°C. An in-line high-capacity cartridge flow pump was attached to the inflow line to supply fluid through the foundations at a rate of

5 mL/s, needed to ensure turbulent flow conditions in the heat exchange tubing and to overcome potential friction losses. Details of the heat exchange fluid used in these tests are presented in [MCC 11] and [STE 12]. The foundation flow valve and bypass flow valve are critical components for controlling the foundation temperature. To pre-heat the fluid, the bypass valve can be opened while keeping the foundation flow valve closed. During testing, the foundation flow valve and bypass flow valve can be opened and closed using LabView motor control software to supply the fluid at a given flow rate to reach a desired average temperature in the foundation. The temperatures of the fluid entering and exiting the foundation are monitored using pipe-plug thermocouples, which reflect the energy injected into the foundation.

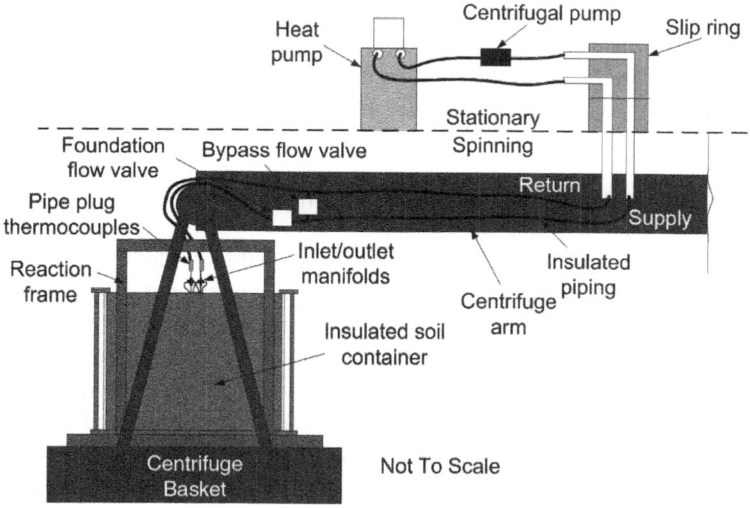

Figure 5.3. *Schematic of the container within the loading frame on the centrifuge showing details of the temperature control system*

5.5. Centrifuge modeling tests for semi-floating foundations

5.5.1. *Soil details*

Soil obtained from the Bonny dam near the Colorado–Kansas border was used in the energy foundation modeling tests presented in this chapter. Information on the compaction curve, shear strength, soil-water retention curve and shear modulus can be obtained from [STE 12]. The liquid and plastic limits of the soil measured according to ASTM D 4318 are 26 and 24, and the fines content of this soil is 84%, so this soil classifies as ML (inorganic silt) according to the Unified Soil Classification System (USCS). The silt has a specific gravity G_s of 2.6. The silt was used in these tests as temperature is not expected to lead to changes in soil-pore water interactions (i.e. diffuse double layer effects) due to its low plasticity, while still behaving like a low-permeability material due to its fines content.

Although a wider range of soil preparation and saturation conditions are under investigation, the tests performed in this study involve soil layers prepared using compaction to permit fast model preparation times and to reach uniform initial unit weight and water content distributions with height at the beginning of the tests. Further, compaction was expected to lead to a stiff soil response that would not lead to significant settlement during centrifuge testing. Foundation A was tested in compacted silt with a gravimetric water content of 13.2% at a dry unit weight of 17.2 kN/m^3. Foundation B was tested in compacted silt with a gravimetric water content of 13.6% at a dry unit weight of 17.4 kN/m^3. A vibratory hammer with a flat-plate adaptor having a width of 75 mm was used to compact the soil beneath and around the foundations to reach lifts with a final thickness of 75 mm. The centrifuge tests were performed on the soil layers in as-compacted (unsaturated) conditions.

5.5.2. *Foundation A: isothermal load tests to failure*

Each foundation shown in Figure 5.2(a) was tested individually after the soil and foundations had returned to ambient conditions from a previous test. The tests were performed by passing fluid with a controlled temperature through the foundations. The foundations were evaluated for the case that there is no building load applied to the top of the foundation, which means that they were free to expand upward during heating. They may also have expanded downward and compressed the soil at the toe, by mobilizing friction between the soil and foundation. The results for the inlet and outlet temperatures for one of the foundations are shown in Figure 5.4(a). Because there was no instrumentation incorporated into foundation B, the inlet temperature was assumed to represent that of the foundation. The average temperature of the soil at different radial locations was then permitted to reach a constant value, as shown in Figure 5.4(b). Next, the foundation was loaded to failure using a constant loading displacement rate of 0.08 mm/min as shown in Figure 5.4(c). Load-settlement curves for energy foundations were obtained by plotting the measured load versus head displacement, as shown in Figure 5.4(d). The load-settlement curves (in prototype scale) for several other energy foundations without a building load during heating are shown in Figure 5.4(d). Because there was no building load during heating, the increase in capacity with temperature noted in this figure can be attributed in part to the increase in radial stresses during heating due to the differential expansion of the foundation and surrounding soil. This increase in radial stresses leads to an increase in ultimate side shear resistance. The foundations that were heated from 15°C to 60°C and then loaded axially to failure experienced an increase in side shear of 40% above that of baseline foundations tested at ambient temperature. A plunging-type failure was noted in the foundations tested under higher temperatures, possibly indicating that the foundation behaved in a more brittle fashion due to the greater lateral stresses induced by expansion of the foundation.

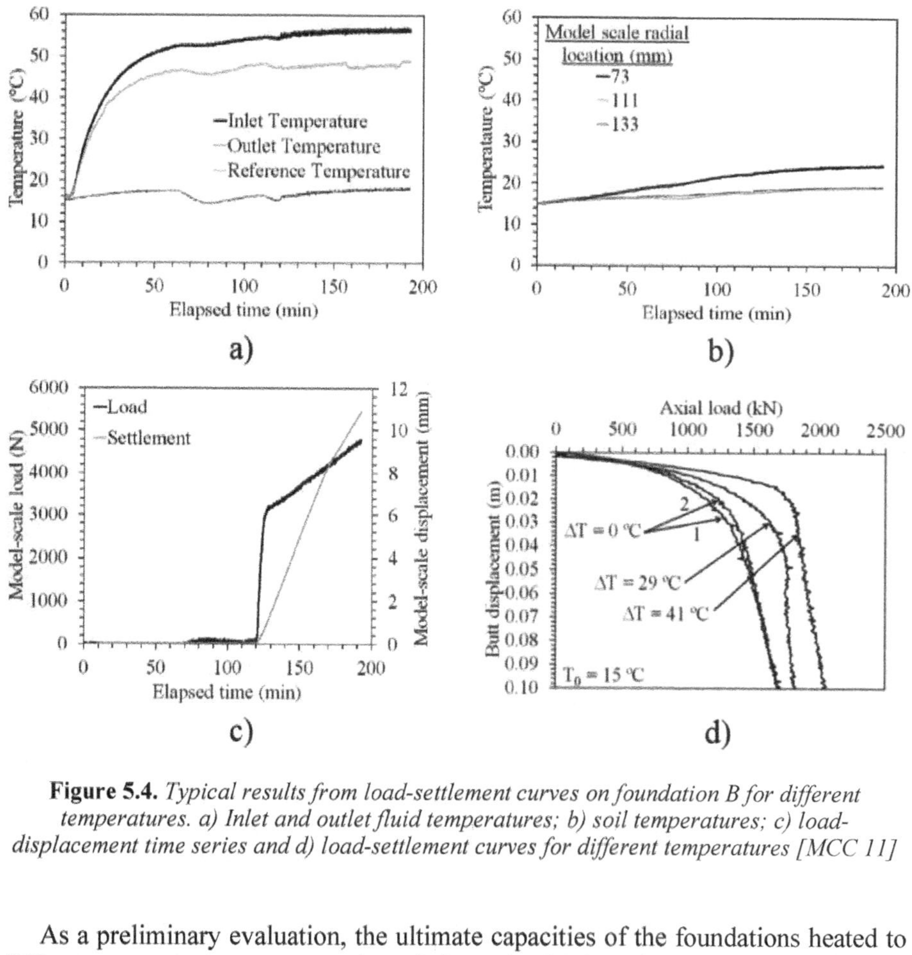

Figure 5.4. *Typical results from load-settlement curves on foundation B for different temperatures. a) Inlet and outlet fluid temperatures; b) soil temperatures; c) load-displacement time series and d) load-settlement curves for different temperatures [MCC 11]*

As a preliminary evaluation, the ultimate capacities of the foundations heated to different temperatures were evaluated from the load-settlement curves shown in Figure 5.4(d) using Davisson's criterion:

$$Q_{ult} = 0.0038m + 0.01D + QL/AE \qquad [5.3]$$

where D is the foundation diameter in prototype scale and QL/AE is the elastic compression of the foundation. The capacity of the two baseline (ΔT = 0) tests was 1,380 kN, while the foundations with changes in temperature of 29°C and 41°C had capacities of 1,700 and 1,820 kN, respectively.

McCartney *et al.* [MCC 10] performed similar tests to those shown in Figure 5.4, but evaluated the behavior of foundations heated under application of a building load, maintained in load-control conditions so that the foundation was still free to expand upward. Application of the building load was observed to cause some consolidation of the soil at the toe of the foundation. After stabilization under the building load, one of the foundations was heated to 50°C (the centrifuge temperature was constant at 15°C) and then loaded to failure. Another foundation was heated to 50°C, cooled down to 20°C and then loaded to failure. The capacity of the test loaded under a temperature of 50°C was 2,150 kN, while the capacity of the foundation loaded after cooling was 1,640 kN. The capacity of the foundation that was heated and cooled was still greater than that of the baseline test (1,380 kN). This is due to consolidation of the soil at the foundation tip, as well as an increase in side shear resistance due to consolidation under the increased radial stresses during heating.

In a load transfer analysis, the mobilization of end bearing with tip displacement can be represented using a Q-z curve. The ordinate of this plot is the normalized end bearing (ratio of mobilized end bearing to ultimate end bearing), and the abscissa is the displacement of the pile toe. Similarly, the mobilization of the side shear with displacement can be represented using a T-z curve. The ordinate of this curve is the normalized side shear (ratio of actual side shear to ultimate side shear), while the abscissa is the relative displacement between the shaft element and surrounding soil. Q-z and T-z curves can be defined using hyperbolic functions, with parameters selected to fit the shapes of the experimental load-settlement curves for the baseline cases [MCC 11]. Additional testing is required to evaluate the impact of temperature on the shape of the Q-z and T-z curves. The non-isothermal shear strength results from [UCH 09] indicate that there is a thermal softening effect on the shear strength of soils, emphasizing the need for research in this area.

The other important inputs for a load-transfer analysis are the ultimate side shear and end bearing capacities. The end bearing is not expected to increase substantially with temperature unless the stiffness of the building system provides a reaction resulting in consolidation of the soil at the toe during foundation expansion. Some increase in end bearing likely does occur due to the downward movement of the lower half of the foundation during heating. Until this is better investigated, the ultimate end bearing can be estimated using conventional bearing capacity analyses, as $Q_b = 9A_bc_u$, where 9 is the bearing capacity factor for deep foundations (i.e. a circular or square cross-section and a depth greater than 2D, where D is the foundation diameter), c_u is the undrained shear strength of the soil under the stress state at the tip of the foundation and A_b is the cross-sectional area of the toe. For the compacted silt evaluated in the centrifuge tests, c_u at the depth of the capacity tests was estimated to be 42 kPa using a value of $c_u/\sigma_v' = 0.265$ and a value of σ_v' estimated using a total unit weight of 17.2 kN/m^3. The estimated value of Q_b is 990 kN.

As an energy foundation expands laterally into the soil during heating, the soil will compress and the interface shear stress will increase. The magnitude of increase in radial stress will depend on the thermal gradient as well as the contrast in linear coefficients of thermal expansion of the foundation and soil. The thermal effects were incorporated into an equation for the drained side shear distribution Q_s by McCartney and Rosenberg [MCC 11], defined as:

$$Q_s = \beta A_s \sigma_v'\left(K_0 + \left(K_p - K_0\right)K_T\right)\tan\phi' \qquad [5.4]$$

where β is an empirical reduction factor representing soil-interface behavior, A_s is the side surface area, σ_v' is the overburden pressure, K_0 is the coefficient of lateral earth pressure at rest (1 – sinϕ'), K_p is the coefficient of passive earth pressure (1 + sinϕ')/(1 – sinϕ') and ϕ' is the drained friction angle (29° for the compacted silt). K_T represents mobilization of lateral earth pressure during heating, defined as:

$$K_T = -\kappa\alpha_c\Delta T\left[(D/2)/0.02L\right] \qquad [5.5]$$

where κ is an empirical coefficient representing the soil resistance to expansion of the foundation, α_T is the coefficient of thermal expansion of reinforced concrete (assumed to be –8.5 με/°C) and [(*D*/2)/0.02*L*] is a geometric normalizing factor. Equation [5.5] accounts for the impact of radial expansion of the foundation, but does not account for the upward relative movement of the upper half of the foundation during heating. The load-settlement curve from the load-transfer analysis was first fitted to the baseline case, as shown in Figure 5.5. After modifying the T-z and Q-z curves to obtain the correct shape for the load-settlement curve, a value of β of 0.55 gave the best fit for the portion of the curve at small displacements.

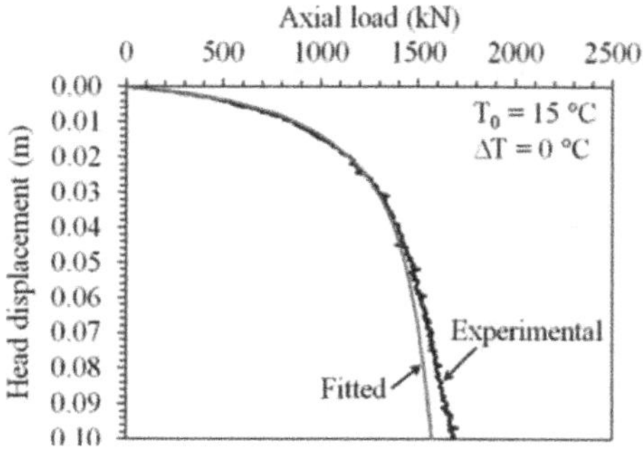

Figure 5.5. *Fitted load transfer analysis for isothermal baseline test to define β [MCC 11]*

The fitted load-settlement curve defined for the foundations heated to a temperature of 50°C is shown in Figure 5.6(a). In this case, a value of κ = 65 along with the same value of β = 0.55 was observed to yield a good fit to the experimental curve. The values of β and κ were then used to predict the load-settlement curve for the foundation heated to a temperature of 60°C, shown in Figure 5.6(b) with a maximum error in prediction of 16%.

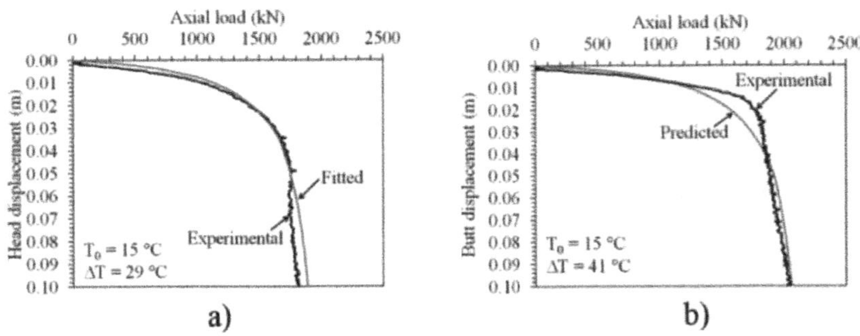

Figure 5.6. *Non-isothermal load transfer analysis results [MCC 11]. a) Fitted analysis to obtain κ and b) prediction using fitted value of κ*

5.5.3. *Foundation B: thermomechanical stress–strain modeling*

The first test performed using foundation B was a load-settlement test under isothermal ambient temperatures. The results from this test are shown in Figure 5.7. The results from this test confirm the assumption from the analysis in the previous section that the side shear distribution is relatively uniform with depth, leading to a decrease in axial strain with depth during loading of the foundation.

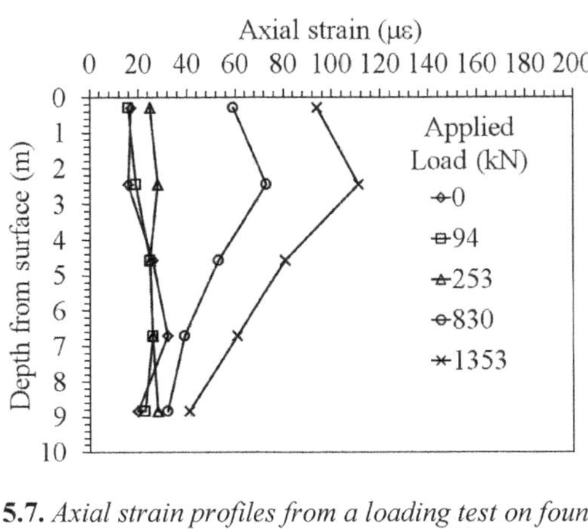

Figure 5.7. *Axial strain profiles from a loading test on foundation B*

A centrifuge test was performed on foundation A (with embedded strain gages and thermocouples) to assess strain and temperature distributions during cyclic heating. The temperature of the heat exchange fluid and the temperatures at different depths in the foundation are shown in Figures 5.8(a) and (b). The foundation was heated to a stable value of 32°C, and then was heated further to 40°C, after which cycles of heating were applied. Temperature profiles shown in Figure 5.8(c) indicate that the foundation temperature was relatively uniform. The temperature of the soil surrounding the foundation measured using the dielectric sensors is shown in Figure 5.8(d). The soil temperature lags behind the foundation temperatures due to the heat flow process. Nonetheless, the thermal axial strains were more sensitive to the foundation temperature than the soil temperature.

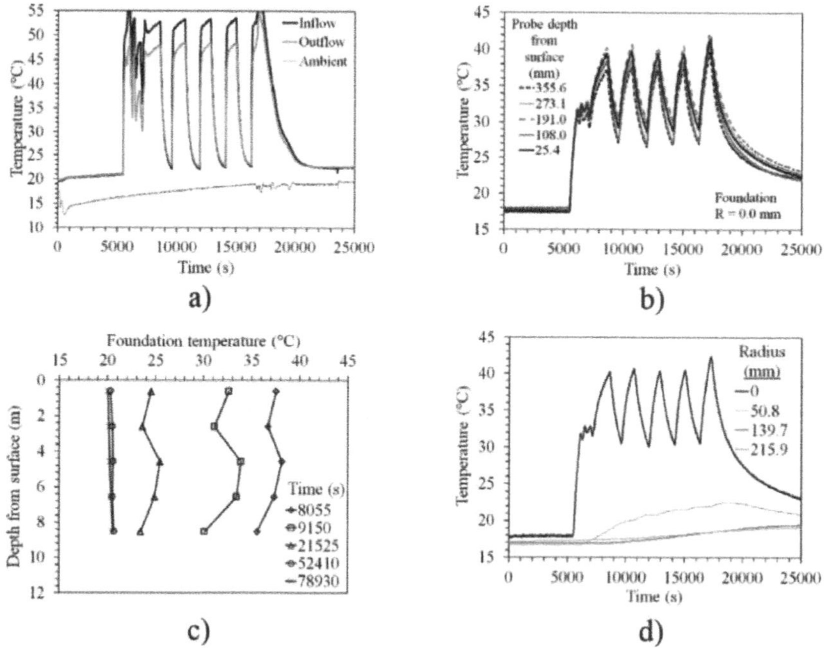

Figure 5.8. *Results from cyclic heating test on foundation B. a) Inlet–outlet temperatures; b) foundation temperatures; c) foundation temperature profiles and d) soil temperatures*

The time series of axial stress applied to the head of the semi-floating foundation is shown in Figure 5.9(a), along with the average foundation temperature. An axial stress of 150 kPa was maintained except during a malfunction of the load control system after 7,000 s. The corresponding mechanical and thermal head displacements are shown in Figure 5.9(b). Although the prototype settlement was nearly 30 mm during the overloading event, this corresponded to a relatively small settlement in model scale (~1 mm). Examination of the head displacement due to temperature changes (accounting for a slight continued consolidation under the building load) indicates a change in the rate of upward expansion of the foundation head after the overloading event. After overloading, the foundation expanded upward by a greater magnitude, likely due to the stiffer soil at the toe of the foundation. After each cooling–heating cycle, the foundation head successively moved downward, due to consolidation and gradual drying of the unsaturated soil around the foundation.

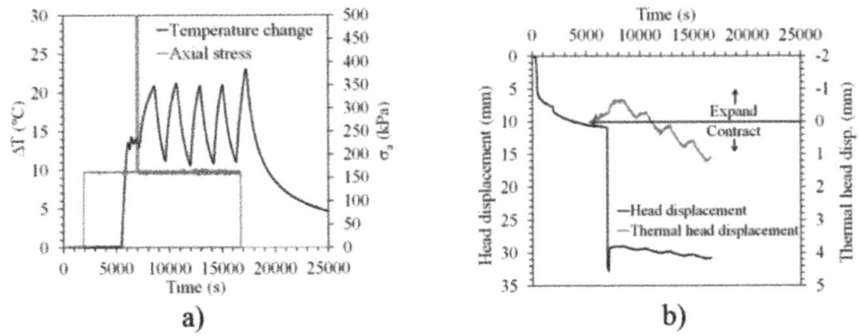

Figure 5.9. *a) Axial stress and temperature and b) thermal and mechanical axial displacement*

The thermal strains during the heating process were defined by zeroing the strain readings at the beginning of heating and applying the thermal correction factors obtained from 1-g tests. The thermal axial strains shown in Figure 5.10(a) indicate consistently negative (expansive) strains in the foundation during heating, which followed the same trends as the imposed foundation temperatures. The magnitude of the thermal axial strains was consistently lower than the free expansion strain of the foundation $\varepsilon_{T,free}$ defined using equation [5.1]. The overloading event led to a small strain in the foundation that was recoverable. The centrifuge was stopped in the middle of the final cooling period, leading to spurious strain values. Profiles of axial strain during the initial heating phase are shown in Figure 5.10(b). The largest strains are observed near the foundation head as it is permitted to expand freely under the constant applied load. The smallest strains occur near the center of the foundation, as it is able to expand both upward and downward in the soil at the tip.

The thermal axial displacement profiles obtained by subtracting the integrated strain profile from the measured thermal head displacements are shown in Figure 5.11(a), with negative displacements for upward movement. The location of the zero thermal displacement is referred to as the null-point [KNE 11]. The null point moves downward as the foundation expands and the soil at the toe stiffens, especially after the overloading event (after ΔT of 14°C). Thermal stress profiles calculated using equation [5.2] are presented in Figure 5.11(b). The maximum thermal stress is located near the middle of the foundation. The soil at the toe provided sufficient restraint that thermal stresses were generated throughout the lower part of the foundation.

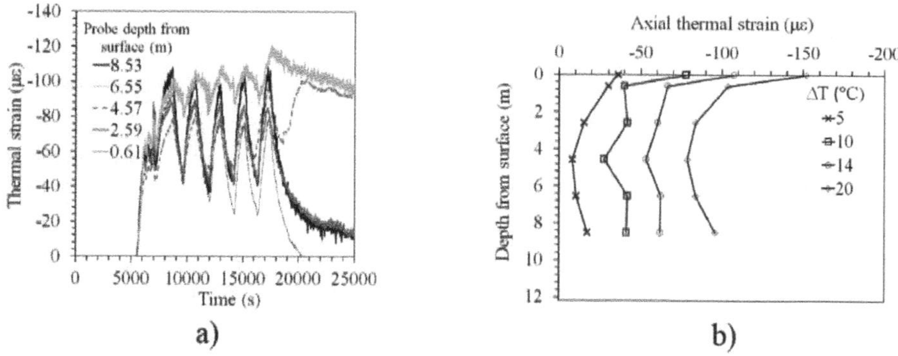

Figure 5.10. *Thermal axial strains. a) Time series and b) profile*

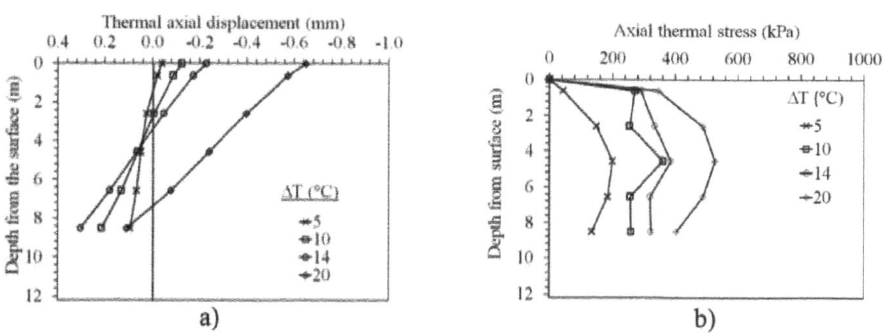

Figure 5.11. *Foundation B results. a) Thermal axial displacement and b) thermal axial stress*

5.6. Conclusions

The results presented in this chapter indicate that centrifuge modeling can be used to define useful parameters for soil–structure interaction analyses related to energy foundations. The thermal axial strain, displacement and stress profiles generated in a semi-floating energy foundation in unsaturated silt reveal the importance of end-restraint boundary conditions on the behavior of semi-floating energy foundations. Slight changes in foundation behavior were noted during cycles of heating and cooling, and the mechanisms for this effect should be studied further.

5.7. Acknowledgments

Financial support from the National Science Foundation grant CMMI 0928159 is greatly appreciated. The views in this chapter are those of the authors alone.

5.8. Bibliography

[AMA 12] AMATYA B., SOGA K., BOURNE-WEBB P., *et al.*, "Thermo-mechanical behaviour of energy piles", *Géotechnique*, vol. 62, no. 6, pp. 503–519, 2012.

[BOU 09] BOURNE-WEBB P., AMATYA B., SOGA K., *et al.*, "Energy pile test at Lambeth College, London: geotechnical and thermodynamic aspects of pile response to heat cycles", *Géotechnique*, vol. 59, no. 3, pp. 237–248, 2009.

[BRA 98] BRANDL H., "Energy piles and diaphragm walls for heat transfer from and into the ground", *Deep Foundations on Based and Auger Piles (BAP III), Ghent, Belgium*, Balkema, Rotterdam, pp. 37–60, 19–21 October 1998.

[BRA 06] BRANDL H., "Energy foundations and other thermo-active ground structures", *Géotechnique*, vol. 56, no. 2, pp. 81–122, 2006.

[ENN 01] ENNIGKEIT A., KATZENBACH R., "The double use of piles as foundation and heat exchanging elements", *Proceedings of the 15th International Conference on Soil Mechanics and Geotechnical Engineering*, AA Balkema, Istanbul, Turkey, pp. 893–896, 2001.

[KNE 11] KNELLWOLF C., PERON H., LALOUI L., "Geotechnical analysis of heat exchanger piles", *Journal of Geotechnical and Geoenvironmental Engineering*, vol. 137, no. 12, pp. 890–902, 2011.

[KO 88] KO H., "Summary of the state-of-the-art in centrifuge model testing", in CRAIG W.H., JAMES R.G., SCOFIELD A.N. (eds), *Centrifuges in Soil Mechanics*, Balkema, Rotterdam, pp. 11–28, 1988.

[KRI 04] KRISHNAIAH S., SINGH D., "Centrifuge modelling of heat migration in soils", *International Journal of Physical Modelling in Geotechnics*, vol. 4, no. 3, pp. 39–47, 2004.

[LAL 06] LALOUI L., NUTH M., VULLIET L., "Experimental and numerical investigations of the behaviour of a heat exchanger pile", *International Journal of Numerical and Analytical Methods in Geomechanics*, vol. 30, no. 8, pp. 763–781, 2006.

[LAL 11] LALOUI L., "In-situ testing of heat exchanger pile", *Proceedings of the GeoFrontiers 2011*, ASCE, Dallas, TX, p. 10, 13–16 March 2011 .

[MCC 10] MCCARTNEY J., ROSENBERG J., SULTANOVA A., "Engineering performance of thermo-active foundation systems", in GOSS C., KERRIGAN J., MALAMO J., MCCARRON M., WILTSHIRE R. (eds), *GeoTrends 2010*, ASCE GPP 6, pp. 27–42, 2010.

[MCC 11] MCCARTNEY J., ROSENBERG J., "Impact of heat exchange on side shear in thermo-active foundations", *Proceedings of the GeoFrontiers 2011*, Dallas, TX, ASCE, pp. 488–498, 13–19 March 2011.

[MCC 12] MCCARTNEY J., MURPHY K., "Strain distributions in full-scale energy foundations", *DFI Journal*, vol. 6, no. 2, pp. 28–36, 2012.

[PLA 12] PLASEIED N., Load-transfer analysis of energy foundations, MS Thesis, University of Colorado, Boulder, CO, p. 90, 2012.

[SAV 88] SAVVIDOU C., "Centrifuge modelling of heat transfer in soil", in CORTÉ J.F. (ed.), *Proceedings of the International Conference on Geotechnical Centrifuge Modeling, Centrifuge 88, Paris, France*, Balkema, Rotterdam, pp. 583–591, 1988.

[STE 12] STEWART M., Centrifuge modeling of soil-structure interaction in energy foundations, MS Thesis, University of Colorado, Boulder, CO, p. 110, 2012.

[TAY 95] TAYLOR R., *Geotechnical Centrifuge Technology*, Blackie Academic & Professional, 296 p., 1995.

[UCH 09] UCHAIPICHAT A., KHALILI N., "Experimental investigation of thermo-hydro-mechanical behavior of an unsaturated silt", *Géotechnique*, vol. 59, no. 4, pp. 339–353, 2009.

PART 2

Numerical Modeling of Energy Geostructures

Chapter 6

Alternative Uses of Heat-Exchanger Geostructures

Heat-exchanger geostructures have mainly been developed as heat-exchanger piles in the foundations of buildings. They mostly use the ground as a heat source for a ground source heat-pump system (GSHP) and can thus be classified as geothermal systems. Another way of using energy geostructures is to couple them with a heat source that transfers heat into the ground and then retrieves this energy seasonally; in this case, the geothermal aspect is limited and the heat storage aspect preponderant. The efficiency of these two solutions depends on the nature of the ground, the presence of groundwater flow and the geometry of the foundation.

Although common foundations of large buildings, in the range of hundreds of piles, have a relatively well-known thermal behavior, there are possible alternative uses for heat-exchanger geostructures that differ widely in terms of geometries and heat demand profiles. Two alternatives are described in this chapter: small foundations for bridge piles and tunnel anchors. The first geometry is very compact, whereas the second is linear. The geotechnical aspects of both of these geometries differ from those of building foundations.

The goal of the first alternative use is to access a local heat source in an efficient manner for the purpose of bridge de-icing. Experiments with borehole heat exchangers (BHEs) have demonstrated the feasibility of this alternative use, but cost concerns have prevented its development. Heat-exchanger piles greatly reduce the

---

Chapter written by Fabrice DUPRAY, Thomas MIMOUNI and Lyesse LALOUI.

initial investment required compared to that for BHEs. The specific aspects of the use of heat-exchanger piles for bridge de-icing are addressed in section 6.1. This application uses energy piles in a context in which they may often be affected by groundwater flow. This aspect is addressed in section 6.1 in the presentation of a thermohydromechanical model for an energy pile, including the effect of groundwater flow. The thermal and mechanical values obtained are compared with the design requirements of a bridge.

The second alternative use changes the nature of the heat-exchanger geostructure from a pile to an anchor. Anchors are used during the construction phase of both bored and cut-and-cover tunnels and can be suitably modified for heat exchange. The possibilities and consequences of their use in tunnels are discussed in section 6.2.

6.1. Small, dispersed foundations for deck de-icing

The problem of ice formation on roads has traditionally been addressed with the use of salt to decrease the freezing point of standing water. Unlike snow, the other weather-related winter danger on roads, ice formation is prone to appear in specific locations, such as humid areas and bridges. Thus, a local approach to the problem of ice formation on roads can be envisioned. Using salt on bridges has been found to cause two types of problem, one for the environment and the other for the durability of the structure. One potential technique for reducing the salt used on bridges is to use the resource of shallow geothermal energy, i.e. energy that can be extracted from shallow ground. The origin of this energy is largely unrelated to natural thermal gradients. The energy source can be solar energy gathered through asphalt or energy transported by groundwater flow. Previous full-scale experiments, such as the SERSO project in Switzerland (operational since 1994), have proven that it is possible to store solar energy in the ground during summer and retrieve it during winter, albeit at a high initial cost due to the need for a storage volume constructed specifically for the site [HOP 94].

This chapter discusses the behavior of an energy pile in this context: evaluation of the energy that can be retrieved from either solar thermal energy or groundwater flow and the related geotechnical response. To analyze these aspects, a thermohydromechanical finite element simulation is performed on a case study. The finite element model used was devised for preliminary investigation of the convection effect due to underground water flow in relationship to both heat storage and the mechanical response of the pile. The case study analyzed is a single pile used as a heat exchanger in both heating and cooling regimes. Four cases are studied: one without groundwater flow and with heat storage, one with moderate flow and heat storage, one with moderate flow and without heat storage and one with significant flow and without heat storage. The thermal responses computed for these four cases are compared, as well as the corresponding mechanical behaviors.

6.1.1. *Heat demand and specificities of small foundations*

6.1.1.1. *Winter needs*

Various published papers [LUN 99, LIU 07] have reviewed the only existing handbook for the evaluation of heat requirements for de-icing and snow melting of bridges, the ASHRAE Handbook for HVAC applications [ASH 95], and have concluded that the approach proposed by ASHRAE is correct on a seasonal basis but has limitations with respect to transient behavior. As this study is focused on seasonal behavior, the ASHRAE method is used for the evaluation of heat needs. Based on meteorological data for the period of interest (i.e. temperatures lower than or equal to the set point temperature of the system, generally 0.5°C), a total amount of energy can be calculated considering snow melting and temperature control. The general equation for the surface heating power needed to melt snow is as follows [ASH 95]:

$$q_0 = q_s + q_m + A_r\left(q_e + q_h\right) \qquad [6.1]$$

where q_0 is the total surface heating power needed, q_s is a term related to the snow temperature increase, q_m is a term related to the latent heat of melting, A_r is the free area ratio, and q_e and q_h are loss terms related to evaporation and convection/radiation, respectively. q_s is proportional to the snowfall rate and temperature, whereas q_m is only proportional to the snowfall rate. Evaporation loss terms are more complex, as they involve wind speed and the relative humidity of the atmosphere.

The heating power needed to maintain the temperature of the water film at a set value (generally 0.5°C) in the absence of snow is identical to the term q_h in equation [6.1] and has been evaluated empirically as follows [ASH 95]:

$$q_h = 64.74\cdot\left(0.0125\cdot V + 0.055\right)\cdot\left(T_f - T_a\right) \qquad [6.2]$$

where V is the wind speed in kilometer per hour, T_f is the film temperature and T_a is the air temperature in degrees Celsius. The resulting value is of the order of 60–200 kWh/m^2/year depending on local conditions.

6.1.1.2. *Possible solar energy retrieval*

The other part of the problem is evaluating the energy that is available and renewable on a yearly basis. As mentioned previously, two main sources of energy can be identified: groundwater flow, assumed to be constant in temperature, and solar energy, ideally retrieved from the bridge itself and stored seasonally. The solar energy retrievable through piping located under asphalt has been found to be on the order of 20% of the total irradiated energy [BOB 13], which is sufficient to cover winter needs in Switzerland, as the total irradiated energy on horizontal surfaces is generally greater than 1,000 kWh/m^2/year.

6.1.1.3. *Possible groundwater flows*

Natural groundwater flow is known to be detrimental to the operation of solar energy storage systems [PAH 07]. The thermal efficiency of a heat storage system is based on its ability to limit thermal losses. These losses depend on ground parameters (thermal conductivity, volumetric heat capacity and permeability) and hydraulic ground conditions (degree of saturation and water pressure gradient). Thus, seasonal energy storage requires particular ground conditions, such as little or no natural groundwater flow. The results of a numerical study by van Meurs [VME 85] concerning a porous medium with homogeneous hydraulic properties indicate that a heat storage unit requires a protecting hydraulic screen if the groundwater flow exceeds 0.05 m/d [NOR 00]. Although thermal conductivity is necessary to spread heat from the piles to the ground, heat losses from the storage area increase with increasing thermal conductivity [SCH 06].

6.1.2. *Modeling of the pile*

6.1.2.1. *Geotechnical context*

A context was devised in this study based on a possible site for a thermo-active bridge in the Rhône Valley of Switzerland. A complete geotechnical study of the actual site was performed and yielded valuable information concerning the nature and geotechnical characteristics of the soil layers encountered. Knowledge about the nature of the soils permits the evaluation of the soils' hydraulic conductivity and their thermal characteristics in the absence of actual measurements. To properly identify and analyze the effects of convection in the process, only the main layer for the length of the piles was considered. This layer is composed of coarse glacial and alluvial deposits. Although these deposits are well graded, their hydraulic conductivity is high at 3.5×10^{-4} m/s (at 20°C). Due to the hydraulic conditions, significant groundwater flow could occur in this layer, which is between 7 and 20 m thick and is constantly saturated.

6.1.2.2. *Geometry of the problem*

During the planning for a bridge recently built in the area, geotechnical investigations were performed and gave indications regarding the soil properties and pile dimensions. The objective of the study is to determine whether the dimensions and number of piles as designed could also be sufficient for the de-icing of the bridge. The pile length varied between 20 and 38 m. The average length of 23 m was chosen as the pile length used in the simulation. The pile diameter chosen was 1 m. A view of the model, including dimensions, is provided in Figure 6.1. Because of the simulation of groundwater flow in non-isothermal conditions, care was taken in the dimensions of the model to prevent cooler water from reaching the boundary during the simulation, which explains the reason for choosing the length of 500 m downward from the pile in the direction of flow. This separation between the cooler water and the boundary eliminates the practical problem encountered with Dirichlet or Neumann boundary conditions that cannot evacuate heat as a function of water flow.

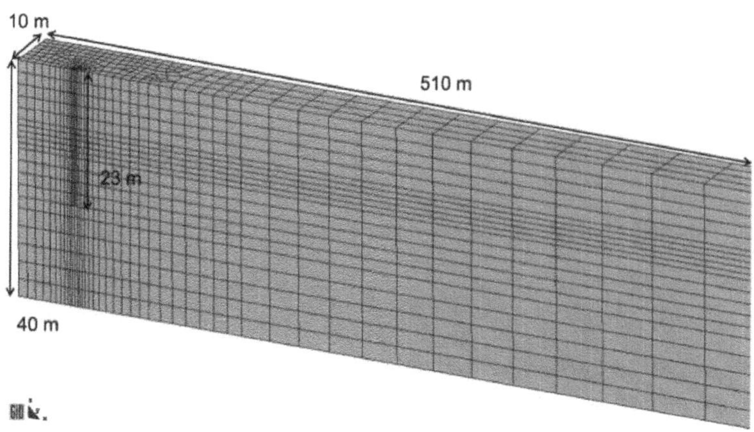

Figure 6.1. *View of the model, including dimensions*

6.1.2.3. *Characteristics of the model*

The numerical simulations were run with the FEM code *Lagamine* [CHA 87, COL 03]. The finite element mesh is composed of 8,022 nodes and 6,340 hexahedral linear elements. A plane of symmetry is used to reduce the model size, and the shape of the pile is hexagonal. The initial pore water pressure corresponds to the hydrostatic profile with a water table located at the surface. Temperature is imposed on all faces except the surface, which is a no-flow boundary. The initial temperature is 12°C throughout the model, which is a typical value for Switzerland. The heat exchange is represented by a heat source/sink in the volume of the pile.

The diffusive model that is used in this chapter is presented in [COL 03]. Only a short description of the principles of the approach is provided here. The equilibrium and balance equations, as well as the water and heat flows, are expressed in the moving current configuration through an Lagrangian-updated formulation. The model uses a phenomenological description that averages properties of the medium (enthalpy and conductivity) depending on the spatial repartition of its constituents. Water is considered compressible, and soil grains are considered incompressible, with a deformable solid matrix. Both water and soil grains are assumed to be affected by thermal dilation. Although expressed for the components of the mixture, the thermal and mechanical parameters of both soil and concrete can be more readily summarized in Table 6.1 for the saturated materials.

Parameters\Materials	Soil	Concrete
Porosity [-]	0.34	0.12
Density [kg/m^3]	2,122	2,408
Hydraulic conductivity 20°C [m/s]	3.5×10^{-4}	9×10^{-9}
Thermal conductivity [W/m/K]	1.59	1.56
Specific heat [J/kg/K]	1,426	1,045
Young's modulus [MPa]	20	3,500
Poisson's ratio [-]	0.3	0.2

Table 6.1. *Thermal and mechanical parameters of the materials involved*

6.1.2.4. *Loading paths*

Three stages corresponding to the three aspects of the problem are necessary to describe the load path. The first aspect is linked to hydraulic behavior. When a water flow is present, the flow is considered to be perpendicular to the bridge (parallel to the canal). Such a water flow is imposed through a constant water pressure difference between the sides of the model, which is allowed to equilibrate throughout the model before the beginning of the mechanical and thermal stages. Two groundwater flow values are evaluated: a high value of 0.3 m/d and a low value of 0.015 m/d.

The second aspect is mechanical loading. Initial stresses due to gravity are introduced in the model by assuming a coefficient of earth pressure at rest of $K_0 = 1$. Loading is then applied to the top of the pile. The individual pile was designed to withstand a load of 2,750 kN, applied as a stress of 3.9 MPa on the top surface of the pile. All subsequent relative values (e.g. displacements and stresses) are based on the reference state at the end of this stage.

The final aspect is the thermal loading. As mentioned previously, two energy sources are considered. In the first case, solar energy absorption is used as the heat renewal source, meaning that heat is injected from the piles to the surrounding soil during the summer months. From June to August, the injection design value is constantly applied, whereas May and September are used as ramp-up and ramp-down periods, respectively. April and October are not used for heat exchange in the solar energy case. Both energy source strategies are illustrated in Figure 6.2.

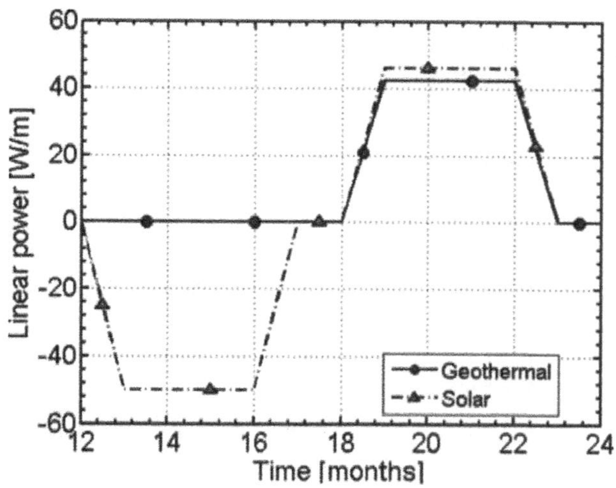

Figure 6.2. *Description of the thermal extraction (positive) and injection (negative) expressed as the linear power imposed along the pile for the case of low water flow*

For both the cases, a similar strategy is used for heat extraction from the ground to the bridge. From December to February, the extraction design value is applied constantly, whereas November and March are used as ramp-up and ramp-down periods, respectively. With underground water flow (or geothermal renewal), no heat exchange is imposed during the rest of the year. The procedure is repeated for 4 years.

The goal of the simulation study was to assess the potential of such a solution for bridge de-icing. Thus, all design values were modified to extract the maximum of energy while maintaining a sustainable level of temperature during the 4 years. Therefore, the extraction design value must follow two criteria: it must not freeze the concrete or soil, for geotechnical reasons, and it must not exceed the thermal reload (all sources combined).

6.1.3. *Results and analysis*

6.1.3.1. *Base case: low flow with solar energy storage*

The groundwater flow for this case is set at 0.015 m/d, which is lower than the value used as a reference for determining the need for a screen in an underground thermal storage area [NOR 00]. The fact that a single pile, and thus a smaller volume, is considered is accounted for in this choice. The highest design extraction value achieved is found to be 46 W/m, which lies in the observed range of 30–70 W/m [PAH 07]. The corresponding injection design value that allowed for sustainable use of the heat-exchanger pile in the long term is found to be 50 W/m. Figure 6.3 provides the temperature profile in the pile at different times during the first year.

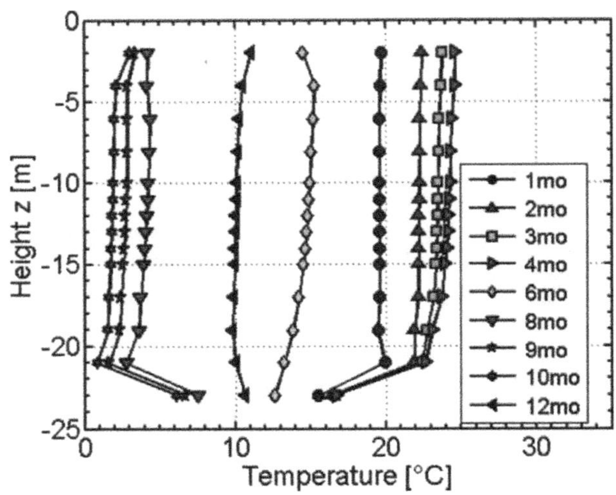

Figure 6.3. *Base case temperature evolution in the pile during the first year*

The effect of this range of temperatures on the mechanical behavior of the pile can be evaluated through the evolution of stresses over time during the period considered. Thermoelastic behavior is considered. Three points are monitored, and the evolution of the average of the thermally induced stresses at the three integration points is provided in Figure 6.4. The evolution is fairly homogeneous throughout the pile, although the maximum thermal stresses are observed in the middle of the pile. The thermal stress magnitude of 0.2 MPa corresponds to 15 kPa/°C, which is logical in the case of low-restraint end conditions observed in an isolated pile, whereas a completely restrained pile would exhibit values of approximately 300 kPa/°C [AMA 12]. These values do not cause concern for the structural integrity of the pile, which is loaded to 3.9 MPa and can be designed to handle the difference. The other aspect of the structure's behavior is linked to displacements especially observed at the surface and transferred to the superstructure. Figure 6.5 shows the thermally induced displacements at the surface and at three other points in the pile. An amplitude of 2.5 mm is observed, which is well within the acceptable displacements of a bridge deck, with the possible exception of short spans (less than 10 m) in cases with the strictest deflection requirements, on the order of *L*/5,000, where *L* is the length of the span.

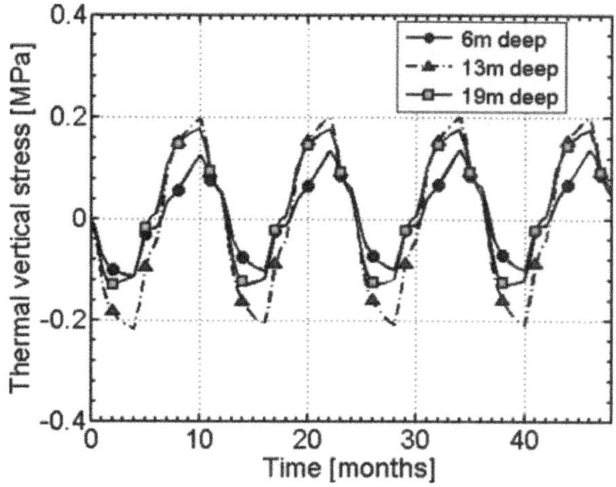

Figure 6.4. *Base case evolution of thermally induced vertical stresses (compression negative) in the pile over time*

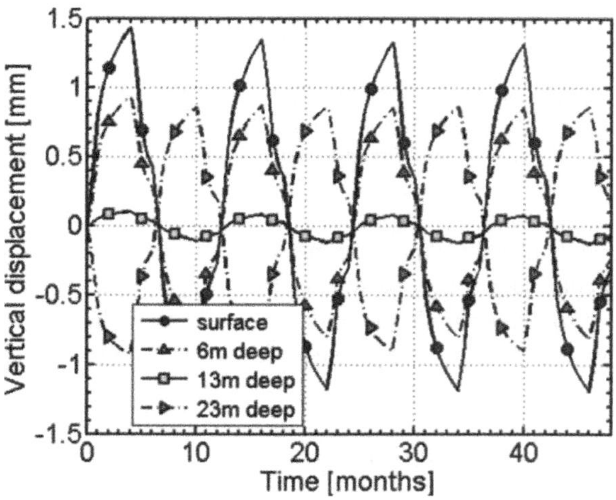

Figure 6.5. *Base case evolution of vertical displacements of certain points in the pile over time*

6.1.3.2. *Case 2: low flow with natural reload*

The design extraction value achieved for this case is found to be only slightly lower than that for the first case, at 42.5 W/m. The small difference is due to the thermal behavior of the system in the base case versus this case, as illustrated in Figure 6.6. Just before the winter season, the temperature difference between the two cases is limited to less than 2°C, whereas it is more than 10°C at the end of the solar reload period. This increased difference at the end of the solar reload period demonstrates the inefficiency of solar energy storage with such a small system compared to a large building foundation. The benefits of solar energy retrieval are limited to two aspects: ensuring a return to the natural temperature even after a demanding winter and controlling the asphalt temperature during hot periods, which enhances its durability and the safety of users.

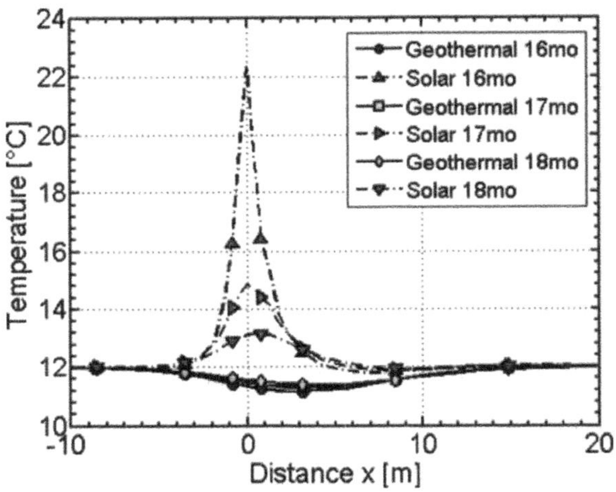

Figure 6.6. *Temperature profiles during the end of the thermal reload period at a depth of 20 m, in the direction of the flow, for the solar and geothermal scenarios (base case and case 2)*

From a mechanical perspective, the major difference lies in the yearly evolution of the displacements. The winter periods of the two cases are comparable, as shown in Figure 6.7. The total amplitude in the geothermal case is limited to 1.3 mm, which is another advantage of the geothermal solution over solar energy use in small foundations.

6.1.3.3. *Case 3: no flow with solar energy storage*

An additional case is tested to determine whether the low efficiency of solar energy storage should be attributed to the groundwater flow or purely to the geometry of the system. In this case, no groundwater flow is present. The design extraction value is found to be equal to that obtained in the base case. Only minimal changes occur in the temperature evolution in the vicinity of the pile, which affect neither the thermal nor the mechanical behavior of the system. These changes are evident in the temperature profiles at a depth of 20 m, shown in Figure 6.8, in which the shift to the right of the full-line curve (low flow) is visible. As a result, the geometry of the system is the main reason for the observed inefficiency, and groundwater flow should be considered as beneficial in any case.

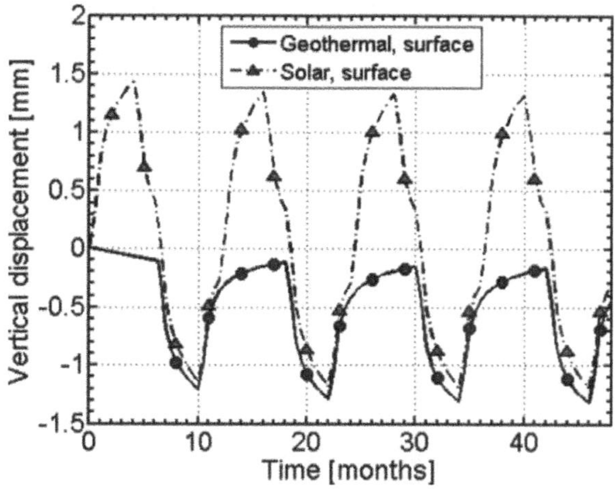

Figure 6.7. *Evolution of vertical displacement of the top of the pile in the solar and geothermal scenarios (base case and case 2)*

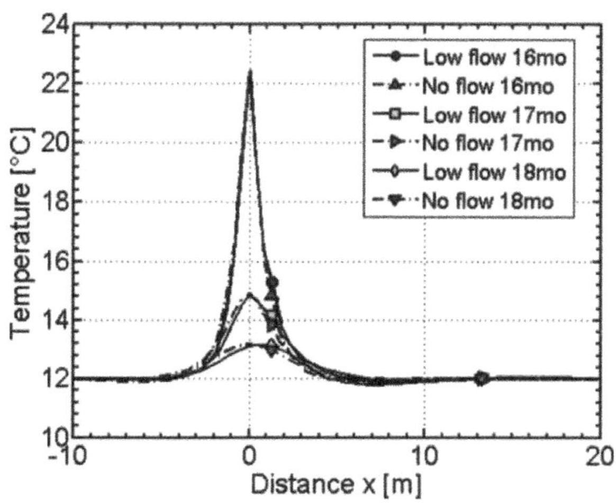

Figure 6.8. *Temperature profiles during the end of the thermal reload period at a depth of 20 m in the direction of the flow*

6.1.3.4. *Case 4: high flow with natural reload*

The evaluation of the potential of isolated (or small groups of) geothermal energy piles highlights the differences between these and a large building foundation. As groundwater flow is beneficial here, a higher water flow of 0.3 m/d is considered in the fourth case. The extraction rate profile shown in Figure 6.2 is used, but a different design extraction value is determined. For this fourth case, a rate of 70 W/m, which is within the range of values that are achieved with conventional heat pumps and energy piles, can be obtained due to the natural thermal reload provided by groundwater flow. A comparison of the evolution of the temperature in the two geothermal cases (cases 2 and 4) is shown in Figure 6.9. The convective effect of the high water flow is clearly visible as the temperature reaches a plateau at 3.5°C during the extraction phase. The thermal reload is also rapid and complete, whereas in the low-flow case, a balanced state over a year is reached only in the fourth year and at a lower temperature level than the natural level. The displacements evolve in a manner similar to that in the other geothermal case, as shown in Figure 6.10. As in the previous case, the displacements are sufficiently small that they do not cause concern.

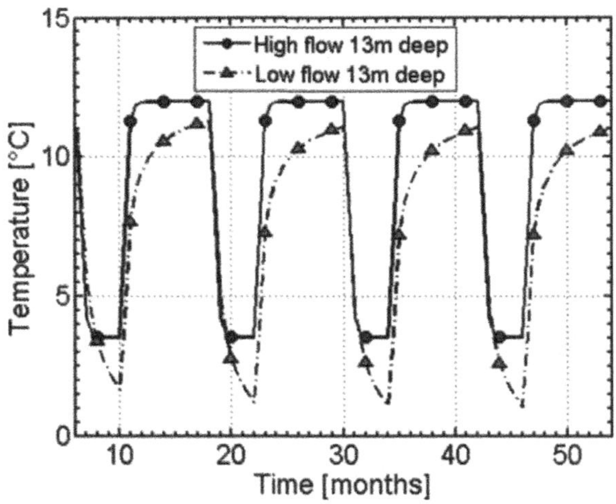

Figure 6.9. *Temperature evolution in the pile for two different flow and extraction rates*

6.1.3.5. *Use for de-icing*

The analysis results presented permit the identification of favorable uses of heat-exchanger piles in the context of bridge de-icing. The total amount of energy that can be retrieved from one pile in each of the four cases leads to a maximum of 5,800 kWh/pile/year in the case of high natural reload through groundwater flow, whereas a less favorable case leads to 3,480 kWh/pile/year. If we consider an average annual heat demand of 100 kWh/m^2, this result translates into a range of 34–58 m^2 of deck per pile, or into a more general unit, 1.5–2.5 m^2 of deck per meter of pile.

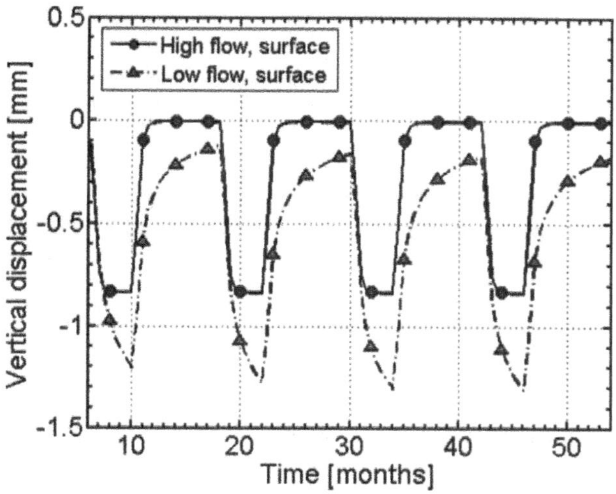

Figure 6.10. *Displacement evolution on the top of the pile for two different flow and extraction rates*

6.2. Heat-exchanger anchors

This section addresses the potential for using anchors or nails for shallow tunnels as heat exchangers with the ground in GSHP systems. Indeed, tunnels that cross deep geologic formations reach higher temperature levels, but heat transport to a final user has been found to be inefficient due to distance [BRA 06]. Shallow tunnels are often built in urban environments, minimizing transport losses, and energy geostructures may be used for heat exchange. Section 6.2.1 describes the technical aspects of the technology and identifies possible users. Section 6.2.2 presents the method of investigation. Finally, the efficiency of heat production through these structures and its mechanical implications are discussed in sections 6.2.3 and 6.2.4.

6.2.1. *Technical aspects and possible users*

Two types of test anchor were installed at a demonstration site in Vienna, Austria [ADA 08]. Coaxial probes were found to be the most efficient configuration because of the relatively narrow shaft of the anchors or nails (Figure 6.11). This technology has two main potential advantages: the ability to produce heat from a renewable source and the ability to cool "hot" tunnels, as defined by Nicholson *et al.* [NIC 13]. The users of the heat produced should be located near the production site, for example in buildings above the tunnels or surface stations.

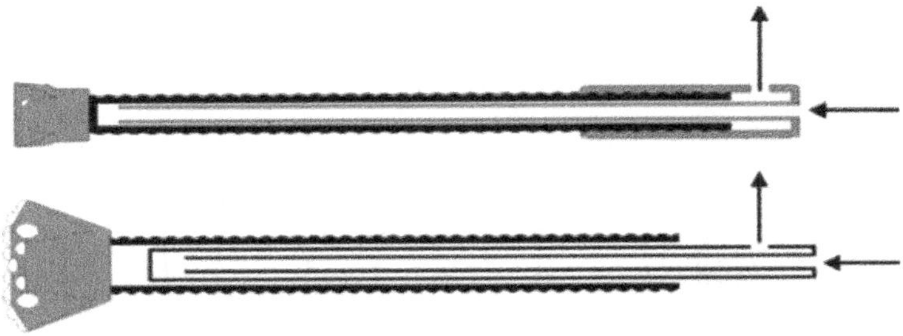

Figure 6.11. *Anchor models used at the Vienna demonstration site: R32N (top) and R51 L (bottom), from [ADA 08]*

6.2.2. *Method of investigation*

Two types of urban tunnels are investigated. The first, a cut-and-cover tunnel (Figure 6.12a) that remains close to the soil surface, is tested in unsaturated conditions and with a varying temperature at the soil surface to account for the thermal influence of the atmosphere on the first few meters of the ground. The anchors used to maintain the diaphragm walls during the excavation period were designed according to the norm SIA-267 [SIA 03].

However, the mechanical implications of heat production through the anchors is neglected as the proximity of the soil surface allows thermal expansion of the soil to develop without inducing significant stresses within the tunnel structure due to the buffering effect of the backfill, which limits interactions between the walls and tunnel.

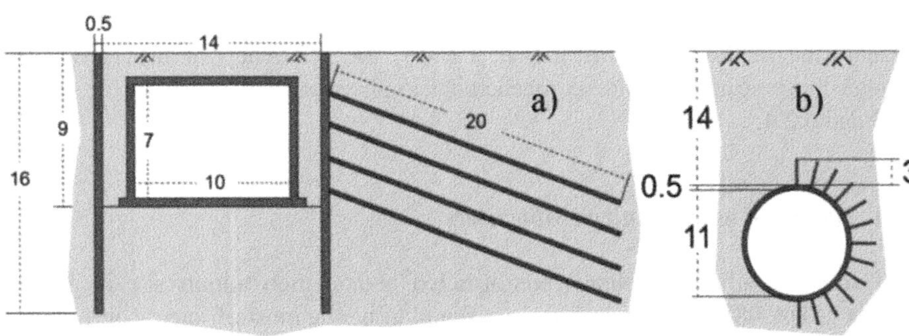

Figure 6.12. *Schematic of the cut-and-cover tunnel a) and the bored tunnel b)*

Second, a bored tunnel (Figure 6.12b) with a circular cross-section is investigated. Due to the depth of this structure, the surface temperature is assumed to be constant over time, and unsaturated conditions are not considered. The mechanical implications of heat production are investigated because of the greater confinement of the structure. A Thermoelastic approach is used to estimate the stress and strain variations in the tunnel lining. Heat production is investigated by means of thermohydromechanical finite element analyses. The analyses are run with the finite element code *Lagamine*. Different soil conditions are tested. Two generic types of soils, slightly permeable (SP) and impermeable (NP), are selected to meet the requirement for seasonal heat storage. Moderately permeable to highly permeable soils allow for significant underground water flow that could compromise heat storage and are therefore outside the scope of the present study. The parameters used to characterize the soils and concrete are grouped as thermal, mechanical and water retention parameters in Table 6.2.

	Thermal Γ_s [W/m/K]; c_s [kJ/kg/K]; ρ_s [kg/m^3]	Mechanical n [–]; k [m^2]	Water retention M [–]; Π [kPa]
NP soil	3.43; 419; 2,700	0.45; 10^{-13}	7/17; 50
SP soil	2.42; 732; 2,700	0.55; 10^{-15}	1/3; 500
Concrete	1.7; 880; 2,500	0.2; 10^{-15}	–;–

Table 6.2. *The thermal, structural and water retention parameters used to represent the porous media*

Finally, different heat production cycles are tested. A simplified heat demand was designed for a building in Lausanne, Switzerland, based on an air temperature time series from the Swiss meteorological service (MeteoSwiss). Next, two types of cycle were designed (Figure 6.13). The first type, Ce, does not include heat injection and thus allows the soil to rest during hot periods. The second type, Cei, includes heat injection during the hot period and is split into two subcategories: Cei,e, which starts the heat production by heat extraction, and Cei,i, which starts with heat injection. Once the shape of the heat production cycle is designed, it is optimized for each possible condition to reach a temperature threshold at a specific location in the ground. The threshold is set at 273 K between the anchors or nails to prevent soil freezing. When heat injection is considered (i.e. for Cei cycles), the injection design value is kept constant for a given tunnel structure to allow for comparisons between the different configurations.

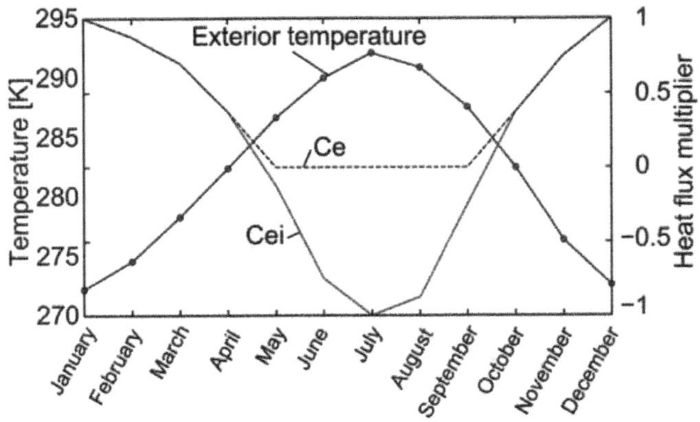

Figure 6.13. *Shape of the heat production cycles with (Ce) and without (Cei) heat injection during the hot periods*

6.2.3. *Optimizing the heat production*

The coefficients applied to the heat flux multipliers are found to be 4 W/m for the cut-and-cover tunnel and 16 W/m for the bored tunnel. The optimization results presented in Tables 6.3 and 6.4 illustrate that injecting heat during hot periods allows for greater extraction during the subsequent cold periods. However, the production gained remains lower than the amount of heat injected due to thermal losses. The efficiency of the heat injection η_{inj} is given by the following equation [MIM 13]:

$$\eta_{inj} = \frac{He(Cei) - He(Ce)}{Hi(Cei)} \qquad [6.3]$$

Cases	*He* (kWh/y/m)	*Hi* (kWh/y/m)	Heat extraction and injection bounds (W/m)
SP – Ce – 0 m	9.63	0	0 to 2.6
SP – Ce – 20 m	7.41	0	0 to 2
SP – Cei – 0 m	14.46	8.41	–4 to 4
SP – Cei – 20 m	12.02	8.41	–4 to 3.33
NP – Ce – 0 m	7.41	0	0 to 2
NP – Cei – 0 m	14.46	8.41	–4 to 4

Table 6.3. *Values of extracted and injected heat for the cut-and-cover tunnel and the bounds of the corresponding heat production and extraction cycles*

Cases	*He* (kWh/y/m)	*Hi* (kWh/y/m)	Heat extraction and injection bounds (W/m)
SP – Ce	45.94	0	0 to 12.4
SP – Cei	57.86	33.62	–16 to 16
NP – Ce	38.68	0	0 to 10.45
NP – Cei	54.92	33.65	–16 to 15.2

Table 6.4. *Values of extracted and injected heat for the bored tunnel and the bounds of the corresponding heat production and extraction cycles*

Heat injection is efficient for the cut-and-cover tunnel: approximately 84% of the injected heat is retrieved in the NP soil, and 55–57% is retrieved in the SP soil. Desaturation of the SP soil does not significantly affect the heat injection efficiency. Conversely, heat injection does not appear to be efficient for the bored tunnel: less than 50% of the heat is retrieved in the NP soil, and only 35% is retrieved in the SP soil (Table 6.5).

	Cases	SP – 20 m	SP – 0 m	NP – 0 m
η_{inj}	Cut-and-cover tunnel	55%	57%	84%
	Bored tunnel	–	35%	48.5%

Table 6.5. *Efficiency of heat injection for the two tunnels*

6.2.4. *Mechanical implications of heat production*

The mechanical implications of heat production were only investigated for the bored tunnel, as explained in section 6.2.2. To obtain a representative initial state (i.e. stresses and strains) of the lining and soil, the excavation was first modeled using the convergence–confinement method [PAN 95]. Because the excavation area is not included in the mesh, equivalent forces are applied along the lining faces to balance the earth pressure. A multiplier (λ) is used to reduce these forces over time. In the present case, λ decreases from 1 to 0.3 in 6 days, which corresponds to a relatively high excavation rate v_λ of 1.35×10^{-6} s^{-1} [CAL 04]. The elements of the lining that were first considered as soil are then given the properties of concrete. Finally, the remaining 30% of the initial natural stress is unloaded over 52 more days. Most of the settlement occurs during the unloading prior to the lining installation. The tunnel crown settles by 0.05 m, whereas the tunnel invert rises by 0.06 m. Once the lining is activated, little settlement is observed, but the lining is put under compression. At the end of the excavation (i.e. $\lambda = 0$), the axial compression developed in the lining is between 1.3 MPa (at the tunnel crown) and 2.0 MPa (at the tunnel invert).

The axial stress variations within the lining during heat production can reach 0.5 MPa when considering heat extraction only, whereas seasonal storage can induce variations of up to 1.5 MPa (Figure 6.14). The variations in the stress observed at the inner and outer faces of the lining oscillate out of phase. During heat extraction (i.e. cooling of the ground), the compressive stress increases along the intrados of the lining and decreases along the extrados (dotted lines in Figure 6.14(a)). These changes in the compressive stresses represent an increase in the load on the lining. Conversely, when heat is injected, the load on the lining is reduced (Figure 6.14(b)). The vertical displacements at the tunnel crown and invert are between 2 and 4 mm. The horizontal displacements of the lining at the top of a virtual invert slab (approximately 2.5 m above the tunnel invert) are between 0.1 and 0.2 mm.

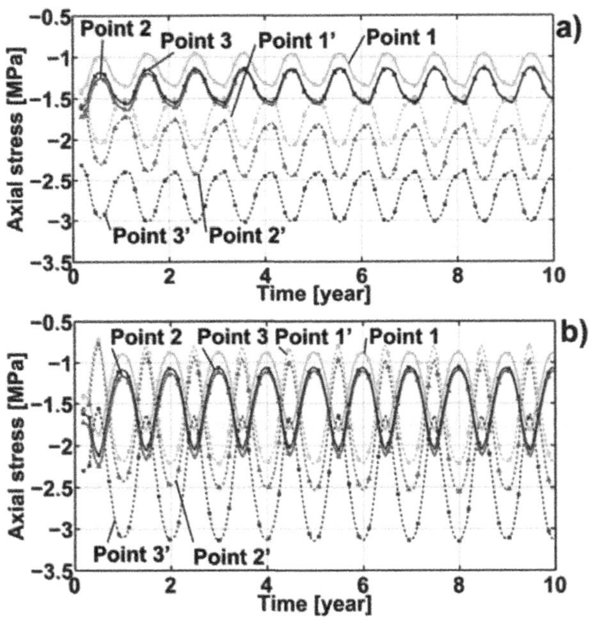

Figure 6.14. *Variations in the axial stress in the lining when considering only heat extraction a) or seasonal heat storage b). Points 1 and 1' are at the tunnel crown (outer and inner faces, respectively), points 2 and 2' are at the middle of the tunnel (outer and inner faces, respectively), and points 3 and 3' are at the tunnel invert (outer and inner faces, respectively)*

6.3. Conclusions

In this chapter, alternative uses and technologies for energy geostructures are evaluated. Small dispersed foundations under bridge piles are found to be usable as energy geostructures for the de-icing of bridges. A comparison between the heat requirements for de-icing and the heat that can be retrieved from the ground or from the asphalt and seasonally stored in the ground is performed. Thermo-hydromechanical simulations, including consideration of groundwater flow, reveal that only the geothermal solution with natural heat reload (from groundwater flow) achieves good efficiency and a satisfying ratio of deck surface de-icing per pile length. The second alternative use studied concerns the geostructure itself, i.e. the behavior of tunnel anchors used for heat exchange. The thermal response of two types of tunnel is studied with external or natural thermal reload in two types of soils. The efficiency of the system is found to be best when using external reload in low-permeability soils with long anchors for a cut-and-cover tunnel or with natural reload for a bored tunnel. In addition, an evaluation of the thermal response of the bored tunnel, including its lining, indicates the need for construction modifications that consider thermally induced dimensional changes.

6.4. Acknowledgments

The authors of this chapter would like to acknowledge the support of the Federal Office of Roads (OFROU), the Swiss Tunnelling Society (FGU), and the Swiss Association of Road and Transport Professionals (VSS).

6.5. Bibliography

[ADA 08] ADAM D., Effizienzsteigerung durch Nutzung der Bodenspeicherung, Presentation, Ringvorlesung Ökologie, TU Wien, 2008.

[AMA 12] AMATYA B., SOGA K., BOURNE-WEBB P.J., *et al.*, "Thermo-mechanical behaviour of energy piles", *Géotechnique*, vol. 62, pp. 503–519, 2012.

[ASH 95] ASHRA E., "Snow melting", *Heating, Ventilating and Air-Conditioning Applications*, vol. 4, pp. 46.1–13, 1995.

[BOB 13] BOBES-JESUS V., PASCUAL-MUÑOZ P., CASTRO-FRESNO D., *et al.*, "Asphalt solar collectors: a literature review", *Applied Energy*, vol. 102, pp. 962–970, 2013.

[BRA 06] BRANDL H., "Energy foundations and other thermo-active ground structures", *Géotechnique*, vol. 56, no. 2, pp. 81–122, 2006.

[CAL 04] CALLARI C., "Coupled numerical analysis of strain localization induced by shallow tunnels in saturated soils", *Computers and Geotechnics*, vol. 31, pp. 193–207, 2004.

[CHA 87] CHARLIER R., Approche unifiée de quelques problèmes non linéaires de mécanique des milieux continus par la méthode des éléments finis, Doctoral Thesis, University of Liège, 1987.

[COL 03] COLLIN F., Couplages thermo-hydro-mécaniques dans les sols et les roches tendres partiellement saturés, Doctoral Thesis, Faculty of Applied Sciences, University of Liege, 2003.

[HOP 94] HOPKIRK R.J., HESS K., EUGSTER W.J., *et al.*, Fact: Federal Roads Office, and Office for Civil Engineering of Bern canton, Technical Report of the Federal Office for Road Engineering/Civil Engineering, Department of the Canton of Bern, 1994.

[LIU 07] LIU X., REES S.J., SPITLER J.D., "Modeling snow melting on heated pavement surfaces. Part I: model development", *Applied Thermal Engineering*, vol. 27, pp. 1115–1124, 2007.

[LUN 99] LUND J.W., "Geothermal snow melting", *Transactions of the Geothermal Research Council*, vol. 23, pp. 127–133, 1999.

[MIM 13] MIMOUNI T., DUPRAY F., MINON S., *et al.*, Heat exchanger anchors for thermo-active tunnels, Report from FGU 2009/002 research project, Federal Roads Office, Bern, 2013.

[NIC 13] NICHOLSON P.D., CHEN A.Q., PILLAI A., *et al.*, "Developments in thermal pile and thermal tunnel linings for city scale GSHP systems", *Proceedings of the 38th Workshop on Geothermal Reservoir Engineering*, Stanford University, CA, 11–13 February 2013.

[NOR 00] NORDELL B., HELLSTRÖM G., "High temperature solar heated seasonal storage system for low temperature heating of buildings", *Solar Energy*, vol. 69, pp. 511–523, 2000.

[PAH 07] PAHUD D., Serso, stockage saisonnier solaire pour le dégivrage d'un pont, Report, Federal Office of Energy, Bern (CH), 2007.

[PAN 95] PANET M., *Le calcul des tunnels par la méthode convergence/confinement*, Presses de l'École Nationale des Ponts et Chaussées, Paris, 1995.

[SCH 06] SCHMIDT T., MANGOLD D., "New steps in seasonal thermal energy storage in Germany", *Ecostock 2006: The 10th International Conference on Thermal Energy Storage*, Pomona, NJ, 31 May–2 June 2006.

[SIA 03] SIA-267, Géotechnique, Swiss Society of Engineers and Architects SIA, Zürich, 2003.

[VME 85] VAN MEURS G.A.M., Seasonal heat storage in the soil, Doctoral Thesis, T. U. Delft, 1985.

Chapter 7

Numerical Analysis of the Bearing Capacity of Thermoactive Piles Under Cyclic Axial Loading

7.1. Introduction

The response of thermoactive piles under combined axial mechanical and thermal cyclic loading has become a major issue in recent years. In general, under static load corresponding to the dead load of the building, a pile foundation is subjected to a compressive stress, which generates settlement along the entire length of the pile. However, in the case of thermoactive piles, the piles are also subjected to thermal stress induced by seasonal temperature variations, thus making a more complex pile behavior that varies throughout the season. If the piles are located in a soil with a substantial groundwater flow (>35 m/ year) or in a sandy soil where the groundwater flows away instantly, the surrounding soil will not be affected by thermal volumetric variations so the ground temperature equilibrium will remain constant [FRO 99, RIE 07]. As a result, only the piles undergo thermal variations and the analysis is limited to the mechanical interaction between soil and pile at the soil–pile interface. Despite the successful number of thermoactive pile installations in Europe and none of these piles experiencing a detrimental effect, a better understanding of the seasonal temperature-induced change in mechanical behavior of a pile is required to optimize the geotechnical design of such piles.

Chapter written by Maria E. SURYATRIYASTUTI, Hussein MROUEH, Sébastien BURLON and Julien HABERT.

Since the ratio between the pile diameter and the pile length is very small, radial movements of the pile induced by temperature variations can be neglected in comparison with thermally induced axial movements of the pile. The response of a thermoactive pile under additional thermal axial loads has shown a strong relationship with the characteristics of the soil–pile interface. This has been confirmed by several *in situ* test results: an important change in mobilized shaft friction and normal stress occurred with the temperature change [BOU 09, LAL 06]. The magnitude of change is related to the degree of axial fixity at the head and the toe of the thermoactive pile foundations [AMA 12, BOU 09]. Although the results from *in situ* tests are the most reliable to determine the response of piles, this method can be quite expensive and time-consuming. Numerical analysis may be another solution to understanding the piles' behavior by modeling interface elements at the soil–pile contact zone using an appropriate behavior law. There are two modeling approaches: the first one uses a load transfer method in a simplified one-dimensional (1D) model and the second one uses a finite element method in a complex three-dimensional (3D) model.

This chapter presents a numerical study of the bearing capacity of thermoactive piles, aiming to ensure the capacity of this type of foundation due to thermal cyclic variations. In section 7.2, a load transfer analysis is conducted to study the response of a pile under a single additional thermal load. Since each year the thermoactive pile undergoes a cycle of temperature variations during the operational time, it is necessary to take into account the cyclic loading effects. A 3D numerical model of a thermoactive pile in cohesionless soil mass is then performed using a constitutive interface behavior law under cyclic loading. The constitutive law, the Modjoin law, can reproduce kinematic hardening under cyclic loads and thus control the cyclic degradation effects [SHA 97, CAO 10]. The influences of different end-connections at the pile–upper structure interface as well as the different service loads applied are discussed.

7.2. Bearing capacity of a pile under an additional thermal load

The load transfer approach, commonly used for the design of piles under axial load, is based on modeling the soil–pile interaction by local laws of mobilized soil–pile resistance (*t-z* law) [FRA 82]. Adopting the method to the design of thermoactive piles under combined axial thermomechanical loads, the calculation consists of decomposing the total deformation ε into an elastic part ε^e and a thermal part ε^{th} (equation [7.1]). Through the compatibility relation, the total deformation is equal to the variation of axial movement Δw at a given depth Δz. Moreover, the axial movement w of the pile at any point is related to the shear stress at that point. Thus, a temperature change in a pile will modify the deformation state and hence the movement of each pile segment, and will consequently induce change in mobilized shaft friction at the soil–pile interface. The differential equation expressing the mechanical equilibrium of the pile is given in equation [7.2]:

$$\varepsilon = \varepsilon^{e} + \varepsilon^{th} \qquad [7.1]$$

$$EA\frac{d^2w}{dz^2} - \pi D\, q_s = 0 \qquad [7.2]$$

where EA is the stiffness of the pile, E is the Young's modulus of the concrete pile, A is the area of the pile, D is the pile diameter and q_s is the mobilized shaft friction at the soil–pile interface. To solve this differential equation, not only is a *t-z* function that governs the relation between the mobilized shaft friction and the soil–pile relative displacement required, but also a boundary condition that controls the bearing stress q_p and tip displacement w_p [FRA 82]. Resolution is completed by an iterative numerical procedure at each shaft segment, seeking the response of the head pile by applying the limit condition at the base.

An example analysis is conducted for a single thermoactive pile with a square section $B = 60$ cm and length $H = 15$ m, founded on a soil mass composed of 12 m of fine soil and 3 m of granular soil. The soil properties are taken from measured data from CETE Nord Picardie. The pile is loaded systematically in two stages: (1) an initial mechanical loading and (2) a single thermal loading. The loading rate at the mechanical stage is taken as 50% of the ultimate load, which is close to the allowable value for the serviceability limit state (SLS) design according to the French standard [AFN 12]. During the second loading stage, a homogeneous temperature variation is imposed over the entire pile, –12°C in the cooling mode and +20°C in the heating mode. In this study, cyclic loading effects cannot be taken into account since only a single thermal loading for each seasonal operation was applied. The temperature variation is translated into thermal axial deformation with the coefficient of thermal expansion of concrete α of 1.2×10^{-5}°C^{-1}. The local law of mobilized soil resistance is governed by the Frank and Zhao *t-z* law [FRA 82]. Three types of pile head condition are considered: (1) a free head pile (zero axial fixity $k = 0$), (2) a restrained head pile (perfectly rigid axial fixity $k = \infty$) and (3) an intermediate pile with a different head axial fixity ($k \in [0, \infty]$) depending on the end-connection to the upper structure.

By this load transfer method, the changes in normal force N and axial displacement w as well as mobilized shaft friction q_s over the length of the pile are obtained. For the sake of clarity, downward movement w and compressive force N are taken to be positive, respecting the convention in soil mechanics. Figure 7.1 shows the typical results obtained for, respectively, the free and restrained thermoactive piles in comparison to the classic pile under a single mechanical load. In a free head pile, cooling the pile in winter induces an additional head settlement together with a reduction in normal forces over the entire pile. In some cases, the decrease in normal forces may even reach a tensile force, depending on the soil resistance around the pile. Heating the pile during summer leads to a pile head heave, accompanied by an increase in normal forces of the pile. Due to the pile expansion, the end-bearing stress strengthens to resist the forces, whereas shaft friction decreases. In a restrained head pile, the results can be interpreted in a similar way. These extreme behaviors allow us to estimate the safety margin of the thermally induced mechanical behavior of the pile. A chart of the global representation of thermoactive pile behavior is finalized, illustrating the variation of thermal axial displacement wth and thermal axial force N^{th} at the pile head relative to the temperature variation ΔT and the head axial fixity k (Figure 7.2) [HAB 12]. This approach is similar to the approach developed by Knellwolf *et al.* [KNE 11].

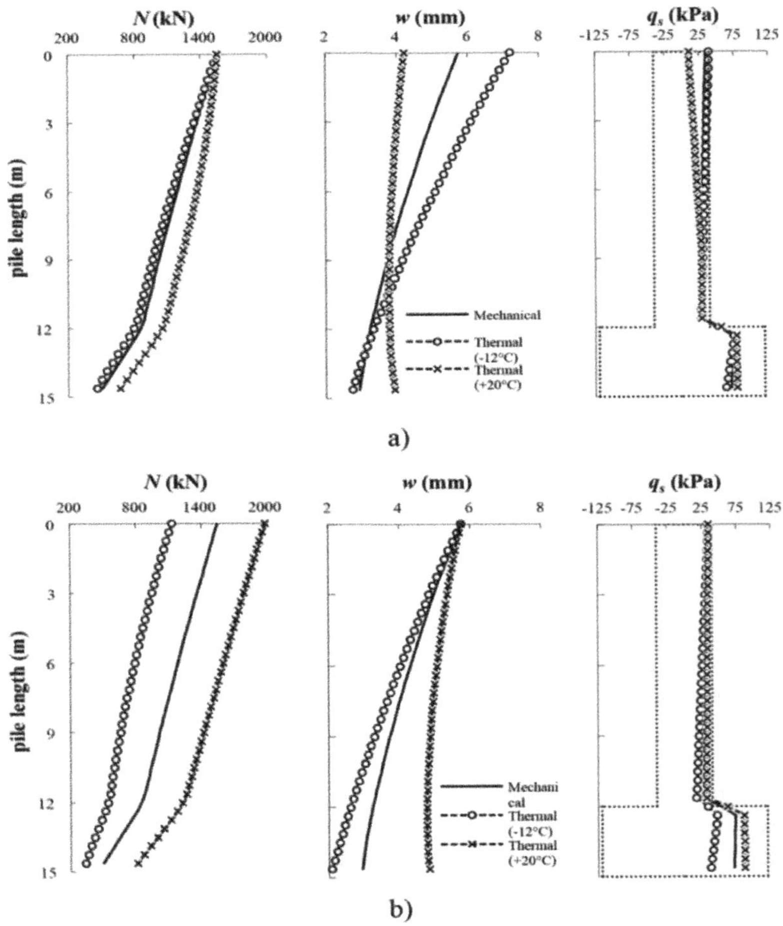

Figure 7.1. *Response of a thermoactive pile under an additional thermal load a) in the free head pile and b) in the restrained head pile*

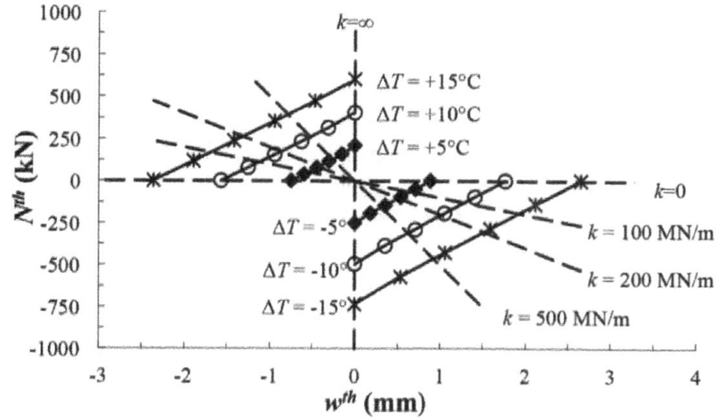

Figure 7.2. *Rigidity chart of a thermoactive pile*

7.3. A constitutive law of soil–pile interface under cyclic loading: the Modjoin law

Generally speaking, the seasonal thermal contraction and dilatation in the thermoactive piles can be equated with a two-way cyclic axial loading. To study the response of the soil–pile interface under two-way cyclic loading, several tests were conducted that resulted in the degradation of shaft resistance during the loading cycles due to dilatancy [POU 89, FAK 97, UES 91]. In cohesionless soil, the degradation of shaft resistance is mainly caused by the maximum shear stress reduction in relation to the absolute slip displacement. Contrary to cohesive soil, it is the decrease in normal effective stress relative to volumetric strain during cyclic shearing that generates the degradation of shaft resistance.

Lemaitre and Chaboche [LEM 85] have defined various phenomena of degradation in cyclic behavior of materials, such as:

– Under imposed non-symmetrical strain, a mean stress relaxation phenomenon appears, indicating a progressive increase/decrease in the stress with cycles.

– Under imposed non-symmetrical stress, the phenomena of strain ratcheting or strain accommodation can be found. Ratcheting refers to the progressive accumulation of plastic strain during cycles, whereas accommodation is the stabilization of plastic strain after certain cycles [LEM 85].

Hence, to accurately estimate the impact of thermal cyclic loading on the thermoactive piles' mechanical behavior, it is essential to use a proper soil–pile interface behavior law that integrates the cyclic degradation effects.

This study uses a constitutive law of interface behavior under cyclic loading, the Modjoin law [SHA 97]. The Modjoin law was established in the laboratory based on the bounding surface concept within an elastoplasticity framework. The Modjoin law incorporates soil non-homogeneity, nonlinearity, postpeak softening–hardening, dilatancy and cyclic degradation effects. Recently, the Modjoin law has been enhanced to control cyclic degradation phenomena, such as stress relaxation, strain ratcheting and strain accommodation [CAO 10].

The elastic part is defined by two parameters: normal stiffness k_n and shear stiffness k_t, which relate the normal stress σ_n to the normal displacement u_n and the shear stress τ to the tangential displacement u_t. The boundary surface f_l and the associated isotropic hardening function $R_{\max}$ are defined by equations [7.3] and [7.4], where $R_{\max}$ is governed by the friction angle φ and the cumulative plastic tangential displacement u_{tr}^p. DR and ADR are the control parameters indicating the amplitude and the rate of isotropic hardening, respectively. The kinematic surface f_c and the associated kinematic hardening function R_c are given in equations [7.5] and [7.6], governed by the parameters γ_c, β_c and λ. γ_c and β_c express the amplitude and the rate of kinematic hardening while λ denotes the plastic multiplier. Finally, the flow rule to reproduce the contracting phase followed by the dilating phase is given in equations [7.7] and [7.8]. This rule depends on the actual plastic tangential displacement u_{tc}^p, the dilation angle ψ_c and the loading rate a_c.

$$f_l = |\tau| + \sigma_n R_{\max} \quad [7.3]$$

$$R_{\max} = \tan\varphi + DR\left(1 - e^{ADR\, u_{tr}^p}\right) \quad [7.4]$$

$$f_c = |\tau - \sigma_n R_c| \quad [7.5]$$

$$dR_c = \lambda\left(\gamma_c \left|R_{\max} - R_c\right|^{\beta_c}\right) \quad [7.6]$$

$$\frac{\partial g}{\partial \sigma_n} = \left(\tan\psi_c - \left|\frac{\tau - \sigma_n R_c}{\sigma_n}\right|\right) e^{-a_c\, u_{tc}^p} \quad [7.7]$$

$$\frac{\partial g}{\partial \tau} = \frac{\tau}{|\tau|} \quad [7.8]$$

Figure 7.3 shows the response of the Modjoin interface compared to a direct shear test performed in CETE Nord Picardie under imposed symmetrical displacement. Depending on the choice of parameters, the Modjoin law can render some cyclic degradation phenomena, as shown in Figure 7.4.

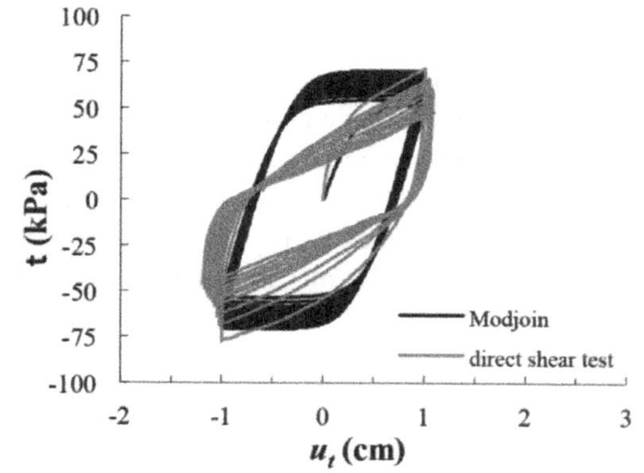

Figure 7.3. *Comparison of interface behavior under imposed symmetrical displacement*

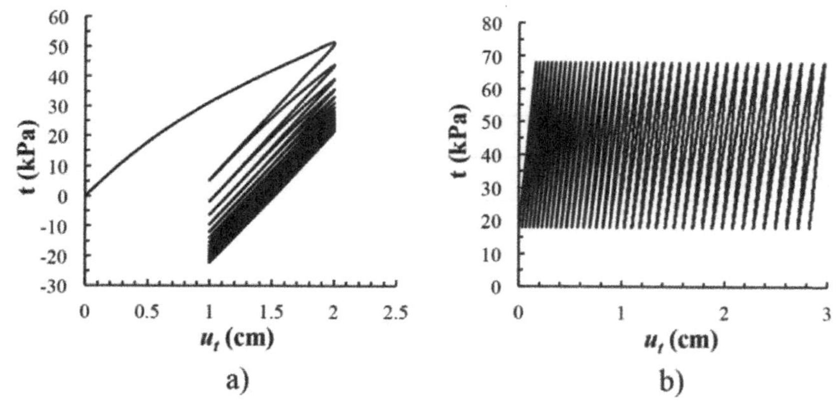

Figure 7.4. *Behavior of the Modjoin interface under non-symmetrical cyclic loading a) stress relaxation phenomenon and b) strain ratcheting phenomenon*

7.4. Numerical analysis of a thermoactive pile under thermal cyclic loading

To study the response of thermoactive piles under combined axial mechanical and thermal cyclic loading, a 3D numerical analysis is conducted using the finite difference code FLAC3D. A single thermoactive pile, B = 60 cm and H = 15 m, is founded on a homogeneous cohesionless soil mass. Both the pile and soil are modeled as solid elements and are assumed to behave in linear thermoelastic conditions. The interface elements are introduced at the zone of contact between the soil and the pile, using the Modjoin law in nonlinear elastoplastic behavior. The

Modjoin interface at the pile base is perfectly bonded to the soil with no separation allowed. The soil properties are taken from the CETE Nord Picardie data for loose sand. The sand is considered to undergo cyclic fatigue, thus parameters *DR* and *ADR* are set at –0.05 in the Modjoin interfaces, referring to a poor soil with cyclic softening behavior. This example of study is a sort of floating pile. Table 7.1 summarizes the properties of the soil, pile and Modjoin interfaces. These properties are assumed not to change with temperature, for which the temperature gradient applied in the pile is in the range ±20°C. Due to the symmetric condition, only one-fourth of the complete domain is modeled. The boundary was set at 15 m (25*B*) of lateral extension of the domain and 30 m (2*H*) of the height of the soil mass, as shown in Figure 7.5.

		Soil	Concrete pile	Modjoin interface
Density	ρ	1,950 kN/m^3	2,500 kN/m^3	–
Bulk modulus Normal stiffness	K k_n	10 MPa –	20 GPa –	– 22 MN/m
Shear modulus Shear stiffness	G k_t	3.75 MPa –	7.5 GPa –	– 8.33 MN/m
Coefficient of thermal expansion	α	5×10^{-6} J/°C	1.25×10^{-5} J/°C	–
Friction angle	φ	–	–	30°
Dilation angle	ψ	–	–	1°

Table 7.1. *Properties of the materials used in the study*

In the first step, the pile is subjected to incremental static loading up to its ultimate load in order to form a load–settlement relationship. In this study, the ultimate load is defined by the applied head load producing a head settlement up to 10% of the pile diameter [BOR 01]. Considering the dead load of the building, the mechanical service load applied to the pile is set at 33% of the ultimate load. This loading stage occurs at $n = 0$ with an index "*mec*". The pile is then subjected to 24 thermal loading cycles, with a temperature gradient of ±10°C from the ground temperature, applied uniformly as thermal axial deformation over the entire pile. Thermal cycles begin with a cooling phase which are followed by a heating phase. The cycles of thermal loading correspond to 12 years of thermal operation. Two extreme end-connections between the head pile and the upper structure are modeled: a flexible connection (the free head pile) and a rigid connection (the restrained head pile). The study focuses on observing three critical responses during a two-way thermal cyclic loading relative to the pile behavior at the mechanical loading stage: (1) the reaction of the pile head to the upper structure, (2) the normal forces distribution in the pile and (3) the mobilized shaft friction at the soil–pile interface.

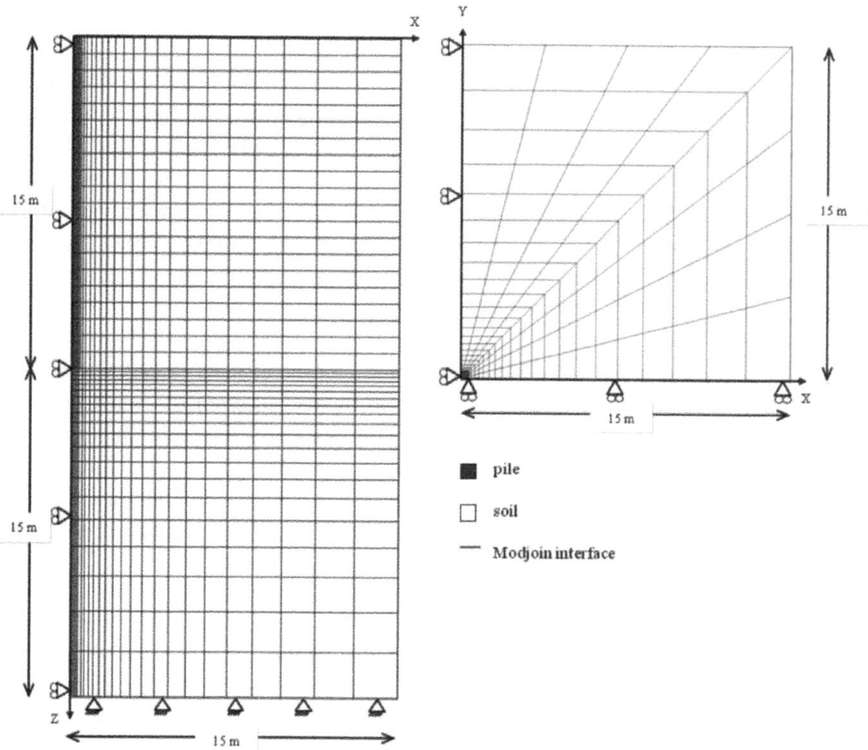

Figure 7.5. *Mesh elements of the 3D model*

7.4.1. *Reaction to the upper structure*

In the free head pile, the temperature variations generate change in head displacement of the order of –5% (heave) to +30% (settlement) from its mechanical settlement. Due to the pile expansion during heating, a pile head heave occurs only in the first five heating cycles and then the pile head settles due to the cyclic degradation of the soil–pile resistance. At the final thermal cycles, the settlement of the pile is about 3 mm deeper than the mechanical settlement.

In the restrained head pile, the head reaction force decreases progressively in the cooling phase from –10% to –25% of its value at mechanical stage. During the heating phase, the first expansion of the pile leads to an additional head reaction force of approximately +7% from its mechanical value and then diminishes during the cycles up to –4% because of the degradation of the soil–pile resistance. Along the 24 thermal cycles, the head reaction force decreases to about 200 kN for a temperature change of 10°C.

7.4.2. *Normal force in the pile*

Figure 7.6 shows the thermally induced normal force over the length of the pile at the beginning and at the end of the cooling–heating cycles. In the free head pile, a reduction in normal force is found at the first cooling cycle. This is due to the pile shortening so that the forces are more mobilized to the shear stress at the soil–pile interface. During the cycles, the change in normal force increases with respect to the degradation of the interface resistance. However, the highest value of normal force variation at the final cycle is about 16% of its value at the mechanical stage (Figure 7.6(a)).

In the restrained head pile, the variation of temperature-induced normal force during the cycles decreases greatly at the pile head, followed by a stiffer bearing resistance due to a bonded interface at the base of the pile with the soil. Figure 7.6(b) shows that the head force decreases about –25% in the cooling phase and –5% in the heating phase at the end of loading cycles. The decrease in normal forces is more significant during the cooling cycles than the heating cycles and can lead to the appearance of tensile force in the pile.

7.4.3. *Mobilized shaft frictions at the soil–pile interface*

Provided that the head is free to move, the mobilized shaft friction shows an opposite response at the upper part and the lower part of the pile. In the cooling phase, the pile shortening results in reducing the tip friction while increasing the mobilized shaft friction. In contrast, the heating phase produces a higher reduction in mobilized friction at the upper part of the pile and thus an increase in tip friction. In this study, the soil–pile interfaces are modeled in cyclic nonlinear behavior using Modjoin law while the soil around the pile is modeled in elastic linear behavior. Hence, the forces induced by cyclic thermal loading are accumulated at the soil-pile interface causing a progressive increase in tangential displacement, so that strain ratcheting phenomenon appears (Figure 7.7(a)). At the lower part of the pile, a lower degradation of the interface resistance was observed due to the presence of the soil underneath the pile that resists the stress mobilized as well.

For the pile with a rigid connection, the mobilized friction gradually decreases along the entire length of the pile. The longer the pile is, the greater the degradation of mobilized friction. Nevertheless, the change in tangential displacement becomes smaller along the pile length. The degradation of temperature-induced mobilized friction is greater in the cooling cycles than in the heating cycles due to the reduction in end-bearing resistance when the pile base lifts during contraction. Figure 7.7(b) shows the response of the soil–pile interface during the cycles, showing a relaxation phenomenon.

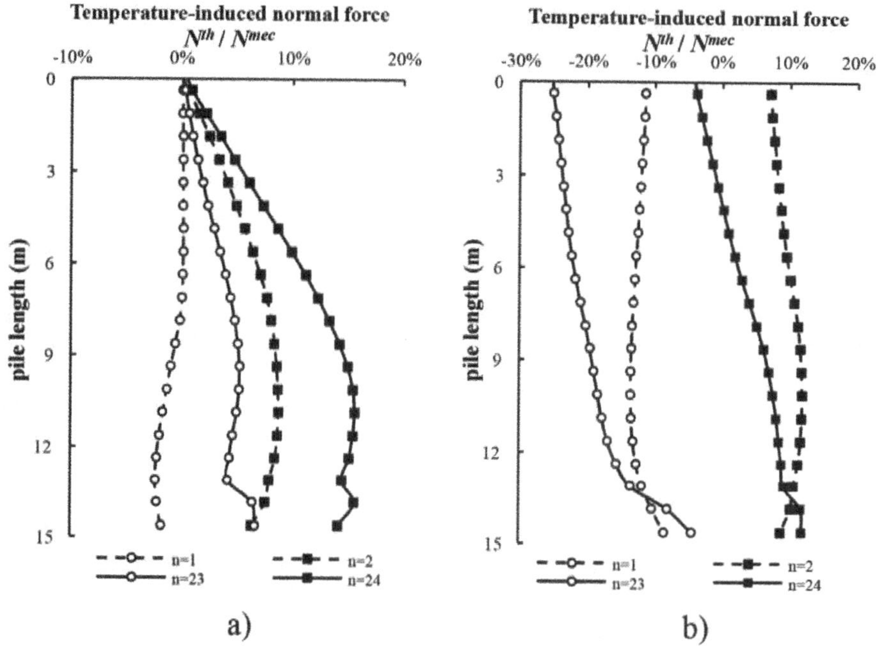

Figure 7.6. *Temperature-induced normal force in the pile a) in the free head pile and b) in the restrained head pile*

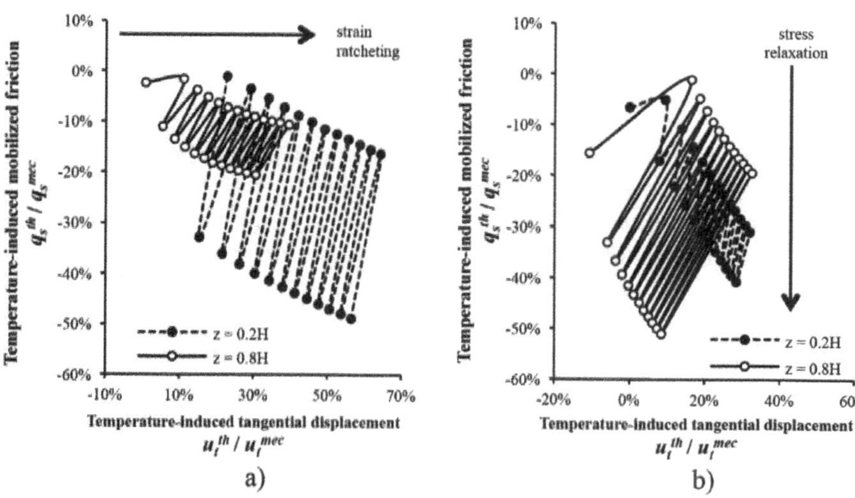

Figure 7.7. *Response of the interface under cyclic temperature variations a) in the free head pile and b) in the restrained head pile*

7.5. Recommendation for real-scale thermoactive piles

7.5.1. *Effect of different loading rates for the applied mechanical load*

According to Eurocode 7, the design of pile foundation can use the ultimate limit state (ULS) and the SLS approaches. Depending on the approach, the loading rate applied at the pile head varies and thus influences the pile dimensioning. Up to the present time, thermoactive piles have been designed in a conservative method by multiplying their safety factor, which indeed causes pile oversizing [KNE 11, BOË 09]. The unresolved issue in the thermoactive pile system is to optimize the design method while ensuring the safety of the foundation under thermal cyclic loads.

The example analyzed in the preceding section has considered a service load of approximately 33% of the ultimate load. Figure 7.8 shows the temperature-induced head settlement for several stages of applied mechanical loading. For thermoactive piles with a small loading rate (i.e. <33% of its ultimate load), the amplitude of thermal head displacement variations during cooling–heating cycles is greater, but the cyclic degradation in head displacement during the cycles is relatively small (i.e. low cyclic softening). In contrast, the higher the loading rate, the smaller the amplitude of settlement–heave variations, but the rate of cyclic softening increases rapidly due to the plasticity of the soil–pile interfaces. For example, after 12 years of thermal operation, the increase in head settlement reaches 75% of its mechanical settlement if the pile is loaded at 42% of its ultimate load. Attention should be paid to the long-term design of the thermoactive piles, considering the loading rate effects on the additional thermal stress produced.

7.5.2. *Effect of thermoactive piles on piled raft foundation*

The building foundation is generally composed of a group of piles connected by a cap or a slab, conventionally known as a piled raft foundation. The use of thermoactive piles for the building thermal system does not necessarily need all of the piles in the foundation. Only a few of the piles in the group can be activated as heat exchangers, according to the energy demand of the building. Since the piles interact within the group, it is important to observe the behavior of thermoactive piles in a piled raft foundation and their influence on the other classic piles.

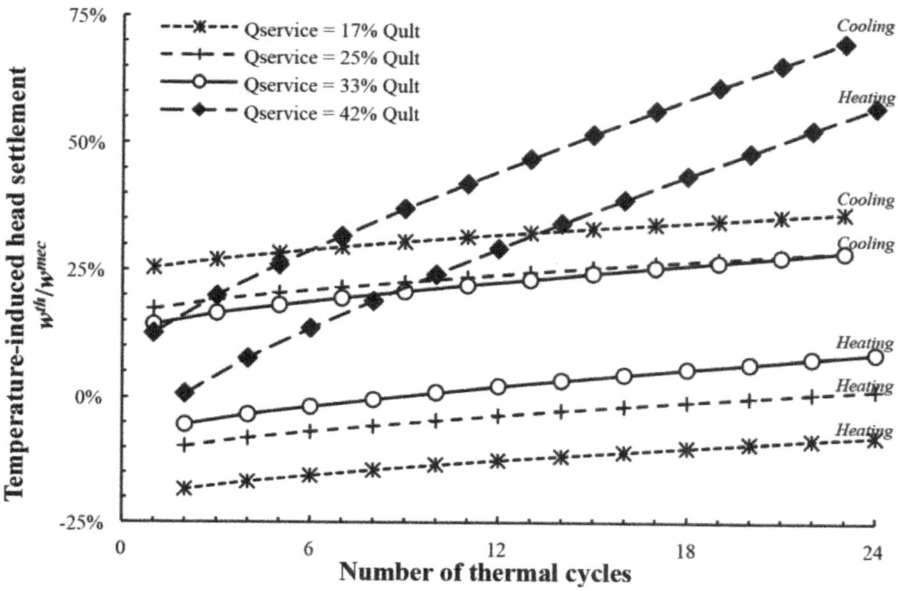

Figure 7.8. *Variations of thermal head displacement in different stages of mechanical loading*

A preliminary numerical study on four thermoactive piles located at the edge of a 3 × 3 pile group was conducted. The piles are identical in size, 60 cm × 60 cm × 15 m. The pile cap is 30 cm thick and 4.8 m wide. The analysis is conducted for only one-fourth of the complete domain due to the symmetry of the problem. Properties and loading stages are identical to the above-described study of a single thermoactive pile. The mesh elements and geometry detail are shown in Figure 7.9. Since the thermoactive piles undergo cyclic contraction–dilatation while the other neighboring piles are subjected only to compressive mechanical load, a high asymmetric concentration of stress occurs in the soil zone between the piles. Figure 7.10 shows the response of each type of pile: one classic pile in the center of the raft, four classic piles at the side of the raft and four thermoactive piles at the edge of the raft. It is shown that the other classic piles consequently undergo the cyclic variation of normal stress due to the interaction between the piles. Further analysis is required to study the influence of the cap rigidity and the position of thermoactive pile in the group on the stress induced by temperature variation.

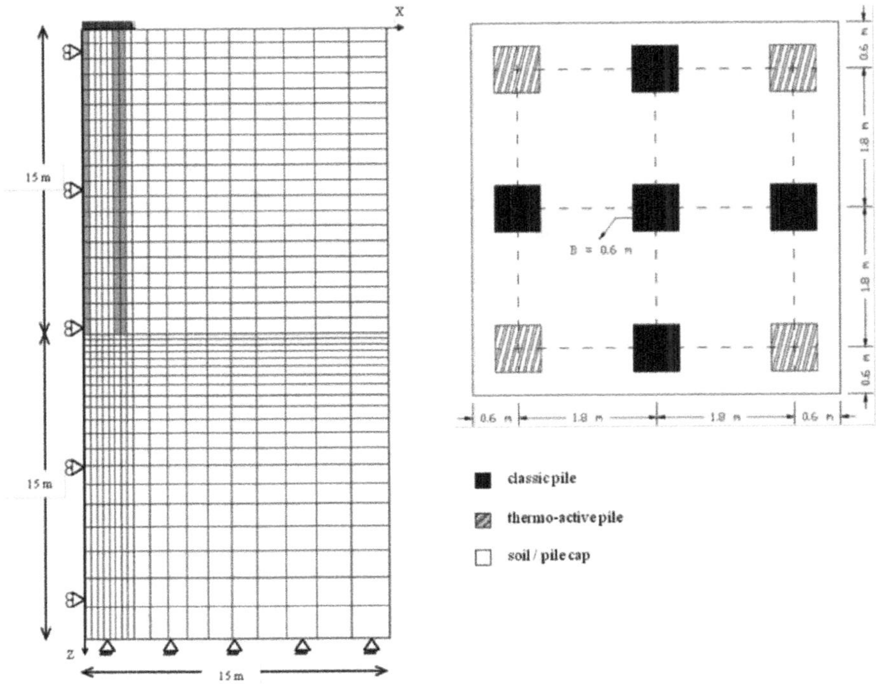

Figure 7.9. *Mesh elements and detail of the raft foundation*

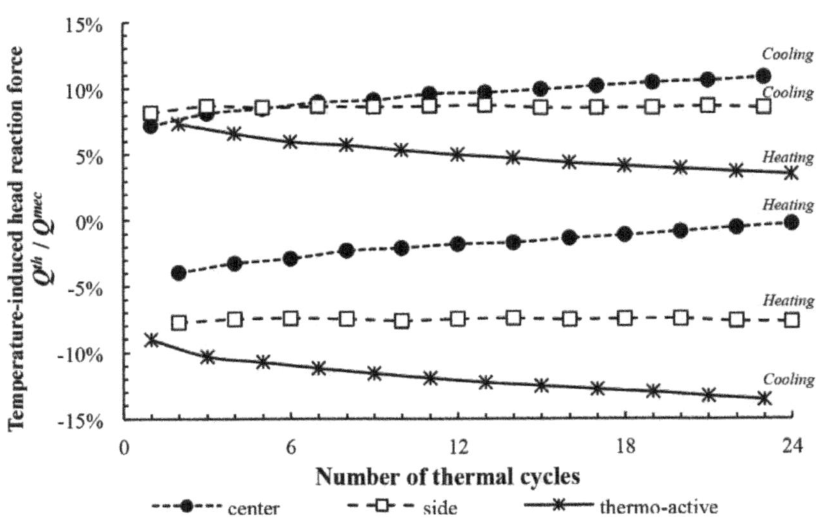

Figure 7.10. *Head reaction force of piles in a 3 × 3 pile group*

7.6. Conclusions

Numerical analyses of the thermoactive piles' behavior under combined axial mechanical and thermal cyclic loads have been conducted in this study. The temperature variations in the pile are translated into thermal axial deformation, applied homogeneously over the entire pile. The analyses begin with a comprehension of the effect of thermal deformation on pile behavior under a single thermal loading. To study the response of thermoactive piles during thermal operation cycles, a constitutive law of the soil–pile interface, called the Modjoin law, is used. The parameters chosen in the model have led to the emergence of cyclic degradation effects: strain ratcheting and stress relaxation. An ongoing project of loading tests on full-scale thermoactive piles will provide real data to calibrate the parameters used in numerical models.

Cyclic temperature variations in thermoactive piles affect the long-term geotechnical capacity in terms of (1) alteration of the head reaction force and head settlement at the pile–upper structure interface and (2) alteration of the mobilized shaft friction and bearing capacity at the soil–pile interface. The magnitude of the additional stress produced is strongly related to the loading rate at the mechanical loading stage. A high loading rate is an advantage in the first thermal cycle, but becomes very risky in the long-term operation of thermal cycles. The values of change in head settlement, normal stress and mobilized shaft friction due to temperature variations can be used to verify (1) the load–settlement capacity, (2) the concrete tensile limit state and (3) the ultimate bearing capacity for both the SLS and the ULS. Thus, a more refined geotechnical design for thermoactive piles foundation can be achieved instead of doubling the safety factor. The interaction factor between piles in a group should not be forgotten as a critical criterion for adequately dimensioning the thermoactive piles.

7.7. Acknowledgments

The work described in this chapter forms part of the research project GECKO (geostructures and hybrid solar panel coupling for optimized energy storage), which is supported by a grant from the French National Research Agency (ANR). It is an industrial project with international collaboration, involving companies and research laboratories in the civil and energy engineering sector: ECOME, BRGM, IFSTTAR, CETE Nord Picardie, LGCgE-Université Lille1, LEMTA-INPL and EPFL. The authors would like to express their gratitude to their partners for their continuous support in this ongoing project.

7.8. Bibliography

[AFN 12] AFNOR, Fondations profondes, Norme française d'application nationale NF P 94-282, 2012.

[AMA 12] AMATYA B.L., SOGA K., BOURNE-WEBB P.J., *et al.*, "Thermo-mechanical behaviour of energy piles", *Géotechnique*, vol. 62, no. 6, pp. 503–519, 2012.

[BOË 09] BOËNNEC O., "Piling on the energy", *GeoDrilling International*, pp. 25–28, GDI March, 2009.

[BOR 01] BOREL S., Comportement et dimensionnement des fondations mixtes, Doctoral Thesis, ENPC, Paris, 2001.

[BOU 09] BOURNE-WEBB P.J., AMATYA B., SOGA K., *et al.*, "Energy pile test at Lambeth College, London: geotechnical and thermodynamic aspects of pile response to heat cycles", *Géotechnique*, vol. 59, no. 3, pp. 237–248, 2009.

[CAO 10] CAO J., Modélisation numérique des problèmes d'interfaes sable-pieu pour les très grands nombres de cycles: Développement d'une méthode de sauts de cycles, Doctoral Thesis, University of Lille 1, 2010.

[FAK 97] FAKHARIAN K., EVGIN E., "Cyclic simple-shear behavior of sand-steel interfaces under constant normal stiffness condition", *Journal of Geotechnical and Geoenvironmental Engineering*, vol. 123, no. 12, pp. 1096–1105, 1997.

[FRA 82] FRANK R., ZHAO S.R., "Estimation par les paramètres pressiométriques de l'enfoncement sous charge axiale de pieux forés dans des sols fins", *Bulletin de Liaison des Laboratoires des Ponts et Chaussées*, vol. 119, pp. 17–24, 1982.

[FRO 99] FROMENTIN A., PAHUD D., LALOUI L., *et al.*, "Pieux échangeurs: Conception et règles de pré-dimensionnement", *Revue française de génie civil*, vol. 3, no. 6, pp. 387–421, 1999.

[HAB 12] HABERT J., BURLON S., "Éléments sur le comportement mécanique des fondations géothermiques", *Actes des Journées Nationales de Géotechnique et de Géologie de l'Ingénieur JNGG 2012*, Bordeaux, pp. 617–624, 4–6 July 2012.

[KNE 11] KNELLWOLF C., PERON H., LALOUI L., "Geotechnical analysis of heat exchanger piles", *Journal of Geotechnical and Geoenvironmental Engineering*, vol. 137, no. 10, pp. 890–902, 2011.

[LAL 06] LALOUI L., NUTH M., VULLIET L., "Experimental and numerical investigations of the behaviour of a heat exchanger pile", *International Journal for Numerical and Analytical Methods in Geomechanics*, vol. 30, no. 8, pp. 763–781, 2006.

[LEM 85] LEMAITRE J., CHABOCHE J.L., *Mécanique des matériaux solides*, Dunod, Paris, 1985.

[POU 89] POULOS H.G., "Cyclic axial loading analysis of piles in sand", *Journal of Geotechincal Engineering*, vol. 115, no. 6, pp. 836–852, 1989.

[RIE 07] RIEDERER P., EVARS G., GOURMEZ D., *et al.*, Conception de fondations géothermiques, CSTB final report ESE/ENR no. 07.044RS, 2007.

[SHA 97] SHAHROUR I., REZAIE F., "An elastoplastic constitutive relation for the soil-structure interface under cyclic loading", *Computers and Geotechnics*, vol. 21, no. 1, pp. 21–39, 1997.

[UES 91] UESUGI M., KISHIDA H., "Discussion of 'cyclic axial loading analysis of piles in sand' by Poulos", *Journal of Geotechnical Engineering*, vol. 117, no. 9, pp. 1435–1457, 1991.

Chapter 8

Energy Geostructures in Unsaturated Soils

8.1. Introduction

As the performance of near-surface geotechnical systems is closely related to water flow arising from environmental interactions, most geotechnical design guides for fill-type systems attempt to minimize the impact of water by using free-draining backfill soils [SAB 97, ELI 01]. However, this can lead to high construction costs, especially in areas where such backfills are not available. In addition to poor drainage, compacted backfills with high clay content are avoided in geotechnical systems because their strength and stiffness tend to decrease with increasing water content (or decreasing suction) during environmental interactions [ZOR 94, MIT 95, ZOR 95]. Silts are avoided for similar reasons and because of their susceptibility to frost heave. However, if these soils remain in unsaturated conditions, these detrimental impacts on the performance of fill-type geotechnical systems with poorly draining backfills may be minimized. It is well known in unsaturated soil mechanics that suction plays an important role in the effective stress state [KHA 98, LU 06, NUT 08]. An increase in effective stress in unsaturated soils can lead to significant improvements in their engineering properties including shear strength and stiffness. A novel way of maintaining unsaturated conditions in poorly draining backfills is to engineer the mechanically stabilized earth (MSE) wall so that thermally induced water flow away from embedded heat exchangers causes drying of the backfill.

An example of a thermally active geotechnical system in unsaturated soil involves placement of heat exchangers into the compacted backfill of an MSE wall.

Chapter written by John S. McCartney, Charles J.R. Coccia, Nahed Alsherif and Melissa A. Stewart.

The heat exchangers can be placed at intermediate lifts between the reinforcements, so that the geosynthetics may serve as permeable pathways for thermally induced water flow. To serve this purpose, the geosynthetic reinforcements may be either permeable woven geogrids or non-woven geotextiles. In this case, the geosynthetics will not only serve their conventional, passive roles (reinforcement, separation and drainage), but will also serve an active role by providing a boundary condition for thermally induced water flow. A schematic of a thermally active MSE wall is shown in Figure 8.1(a). The spacing and location of heat exchangers is not only important for uniform soil–geosynthetic composite improvement, but also to create a thermal gradient between the backfill and atmosphere, which will in turn drive water from the system. A strategy for the routing of heat exchange tubing is shown in Figure 8.1(b).

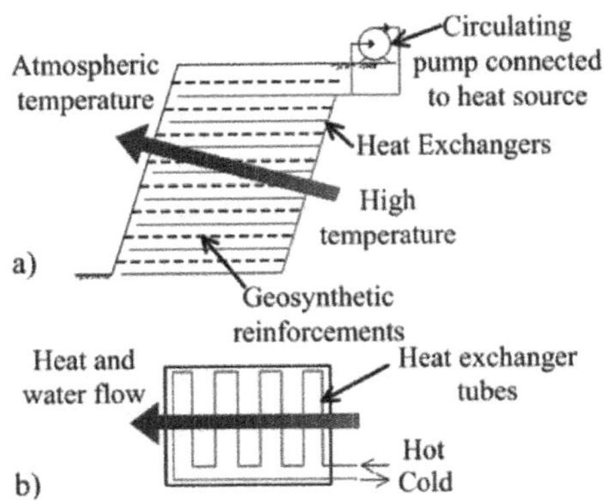

Figure 8.1. *Thermally active MSE wall. a) Elevation section and b) plane view*

Thermally active geotechnical systems have the added benefit of enhancing the energy efficiency of heat pump systems for cooling of buildings or in rejection of excess heat from power plants or industrial facilities. Cooling needs are typically defined in terms of equivalent tons of cooling, defined as the heat rejection in cooling 12 L/min by 5°C [DRB 96]. Buildings typically require 0.82 tons of cooling (4,396 W) per 100 m^2 of floor area [NRE 97] while a 700 MW power plant can require as much as 100,000 tons of cooling [DRB 96]. As heat is rejected through conduction, the inlet fluid temperature, heat-exchanger length/spacing and fluid circulation rate are the important design variables that can be varied to reach a heat rejection goal. The entering water temperature typically ranges from 50°C to 80°C in typical cooling applications [OME 08].

This chapter includes a critical review and summary of data presented in the literature on the different impacts of temperature on the thermohydromechanical behavior of unsaturated soils, as well as a discussion on the implications of temperature on the behavior of geosynthetic reinforcements.

8.2. Thermally induced water flow

Thermally induced water flow in soils is the main driver for maintaining unsaturated conditions in the backfill of thermally active geotechnical systems. A significant body of work has been assembled on this topic in the area of vadose zone hydrology, including a definition of governing equations for water flow in liquid and vapor forms [PHI 57], analytical and numerical solutions [CAR 62a, CAR 62b, TAY 64, MIL 82] and field experiments [MIL 96]. The governing equations have been incorporated into commercial finite element programs, such as VADOSE/W [WIL 94], while the impact of volume change has been evaluated in advanced models [THO 95a, THO 95b, THO 96]. Although analysis of thermally induced water flow under a heat-exchanger boundary condition is important, the role of atmospheric boundary conditions on the flow of heat and water into and out of the system must be considered. The prediction of atmospheric interaction in soil layers has been investigated in landfill cover analyses [MCC 04, ZOR 05, SCA 02, OGO 08].

Two experimental studies of thermally induced flow away from an axisymmetric line heat source in unsaturated soils are shown in Figure 8.2. For a temperature of 90°C applied to a sand-bentonite mixture, Yong and Mohamed [YON 96] noted a decrease in degree of saturation of 0.5 near the heat source after 3 days, with a drying zone of influence of about 40 mm. For a temperature of 70°C applied to sand, Ewen and Thomas [EWE 89] observed a drying over a zone of influence of 20 mm. Although the zones of influence observed in these two examples are small, a greater zone of influence and faster response is expected for silts or low plasticity clays due to their greater hydraulic conductivity over a range of water content.

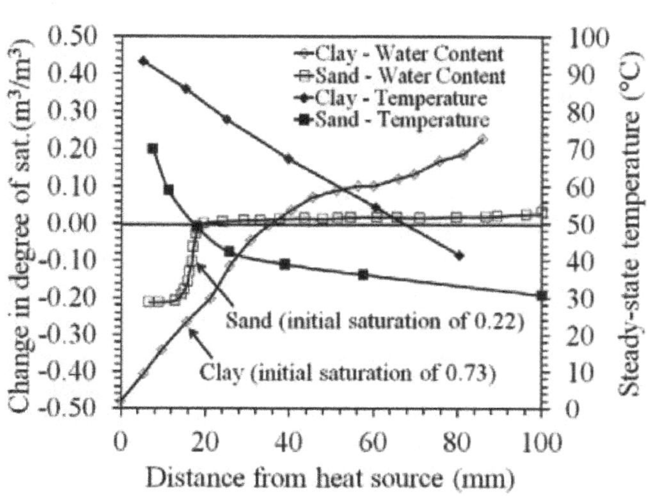

Figure 8.2. *Thermally induced water flow in clay [YON 96] and sand [EWE 89]*

The affected zone and magnitude of thermally induced water flow are functions of the initial degree of saturation (S), hydraulic conductivity for saturated conditions (k_s), thermal conductivity (λ) and porosity (n). Coccia and McCartney [COC 13] observed a greater decrease in saturation at the location of a heat exchanger for silt with lower initial S (Figure 8.3(a)). Furthermore, soils with lower k_s tended to have a smaller zone of drying (Figure 8.3(b)). Thermally induced water flow is also affected by coupled changes in the thermal conductivity of soil, which may decrease by a factor of about 10 as S approaches zero, with a lower limit of 0.25–0.5 W/m°C depending on density and mineralogy [FAR 81]. However, Smits *et al.* [SMI 13] observed nonlinear changes in thermal conductivity for sands during drying under a temperature above 55°C, with a peak value at a saturation of 0.1.

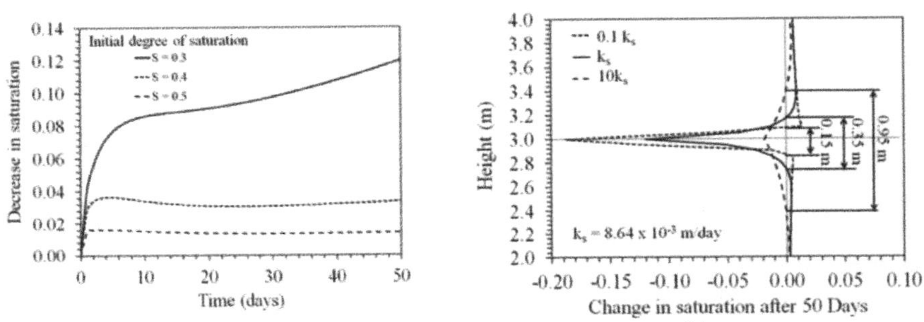

a) b)

Figure 8.3. *a) Decrease in saturation at the location of heat exchange with time. b) Thermal saturation profile as influenced by hydraulic conductivity*

8.3. Thermal volume change in unsaturated soils

An important side effect of soil heating is the possibility for excess pore water pressure generation that may lead to additional flow of water away or toward the heat exchanger, as well as a decrease in undrained shear strength. Campanella and Mitchell [CAM 68] found that positive excess pore water pressures are induced in saturated normally consolidated (NC) soils due to differences in the relative expansion of the water and soil skeleton during heating. Specifically, the coefficient of thermal expansion of pore water is approximately 7–10 times that of most soil particles [MCK 65, MIT 05]. If these excess pore water pressures are permitted to drain, time-dependent, irrecoverable (elastoplastic) contraction or expansion will occur depending on the soil structure and stress history. Saturated NC soils tend to contract plastically during drained heating, and then contract elastically during cooling. Saturated soils with overconsolidation ratios (OCRs) greater than 1.5–3 tend to expand and contract elastically during drained heating and cooling, respectively. This behavior is summarized in Figure 8.4(a) for different soils. Soils with a greater plasticity index show more volume change during heating [SUL 02].

This behavior has been incorporated into constitutive models based on the Cam-Clay model [HUE 90a, HUE 90b, CUI 00, LAL 03, FRA 08, CUI 09].

Different from saturated soils, fewer studies have focused on the thermal volume change behavior of unsaturated soils [SAI 90, SAI 91, SAI 00, ROM 03, ROM 05, FRA 05, TAN 08, TAN 09, UCH 09]. Of these studies, all but two were performed on expansive clays, which would be unsuitable as backfills. Saix *et al.* [SAI 00] performed drained heating tests on clayey silt specimens at a constant suction of 4.5 kPa, and observed plastic contraction during heating. Uchaipichat and Khalili [UCH 09] performed constant water content (undrained) heating tests on compacted silt, and measured a decrease in matric suction of 50% during heating from 25°C to 60°C. The pore water likely expanded and filled more voids, leading to a decrease in suction. A re-evaluation of drained thermal volume change data for unsaturated silt from [UCH 09] as a function of the OCR calculated using the effective stress definition of Khalili and Khabbaz [KHA 98] is shown in Figure 8.4(b). The trends are similar to those for saturated soils, confirming the importance of stress state. However, the reasons for this behavior are different from saturated soils. For the low OCR specimens, it is possible that the change in temperature led to a decrease in the preconsolidation stress and the associated yield function, which may lead to a tendency for an unsaturated soil having a stress state near the yield function to collapse during heating. The trends for unsaturated silt are also confirmed by studies on compacted bentonite [ROM 03, TAN 08].

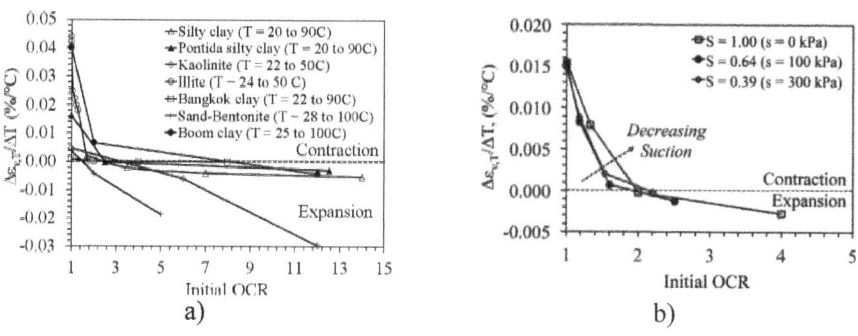

Figure 8.4. *Gradient of drained thermal volumetric strain ($\Delta\varepsilon_{v,T}$) with temperature changes (ΔT) for: a) saturated soils (silty clay [TOW 93], Pontida clay [BAL 88], Kaolinite [CEK 04], illite [PLU 69], Bangkok clay [ABU 07a], sand-bentonite [GRA 01] and Boom clay [SUL 02]) and b) unsaturated compacted silt [UCH 09]*

8.4. Thermal effects on soil strength and stiffness

Temperature generally does not have a significant effect on the engineering properties of most saturated and unsaturated soils. Campanella and Mitchell [CAM 68] studied the compression behavior of saturated soils under different temperatures, and found that the compression index was independent of temperature. This has been confirmed by other studies on saturated soils [DEM 82, GRA 01] and unsaturated soils [SAI 00, UCH 09]. Soils that do show temperature-dependent compression indices are typically expansive clays [DEL 00, ROM 03], which are not suitable for use as backfills. Cekerevac and Laloui [CEK 04] summarized data from the literature indicating that temperature does not have a significant impact on the friction angle. This is supported by studies that found that the critical state line is independent of temperature for saturated soils [FIN 51, KUN 95, GRA 01, ABU 07b, ABU 09] and unsaturated soils [UCH 09]. However, Hueckel *et al.* [HUE 09] hypothesized that the critical state line may be sensitive to the stress history and drainage conditions during heating for some soils, and proposed alternative explanations for data presented in the literature.

Temperature primarily affects the shear strength and stiffness of soils through volume change and changes in the shape of the plastic yield surface. For saturated NC clays, the volumetric contraction after dissipation of thermally induced positive excess pore water pressure has been found to correspond to an increase in strength and stiffness [HOU 85, HUE 92, ABU 07b, ABU 09]. Saturated overconsolidated (OC) soils show a decrease in strength after heating due to volumetric expansion [PAA 67, PLU 69]. During mechanical loading of saturated soils after heating, a lower preconsolidation stress is observed for OC clays. For unsaturated soils, suction causes a hardening effect and an increase in the apparent preconsolidation stress [ALO 90]. Similar to saturated OC soils [ERI 89, TID 89], mechanical loading of soils after heating indicates that the soil has a lower preconsolidation stress [UCH 09]. The data in Figure 8.5(a) from different studies confirm this trend. Saix *et al.* [SAI 00] observed that this decreasing trend may not be general for all temperatures. The effect of a lower preconsolidation stress is typically a more ductile stress–strain curve, as shown in the data from Uchaipichat and Khalili [UCH 09] in Figure 8.5(b). Comparison of the relative impacts of temperature and suction on the shear strength of unsaturated soils indicates that suction has a more significant effect.

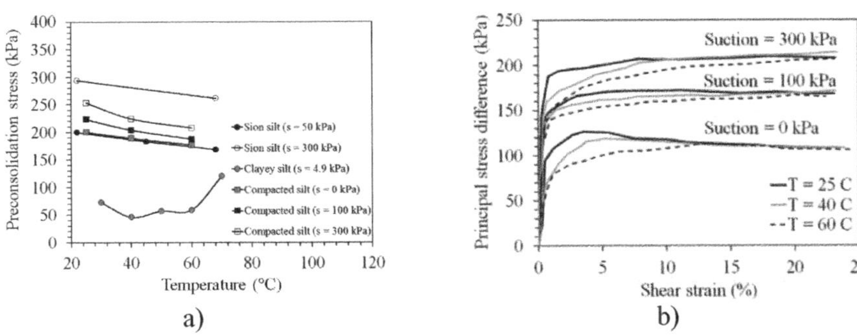

Figure 8.5. *Impact of heating on the a) preconsolidation stress of silts [SAI 00, FRA 05, UCH 09] and b) stress–strain curves of unsaturated silt [UCH 09]*

The trends observed in Figure 8.5 may not hold for the high suction magnitudes that may be encountered in thermally active geotechnical systems. Preliminary work by Alsherif and McCartney [ALS 12] indicates that temperature does not have a significant impact on the shear strength of compacted silts under high suction magnitudes. The stress–strain curves in Figure 8.6(a) for specimens having an initial suction of approximately 100 MPa under the same net mean normal stress but different temperatures were similar, with a relatively brittle shape. The suction–stress characteristic curves derived from failure envelopes for specimens at different initial suctions and different temperatures in Figure 8.6(b) indicate similar behavior.

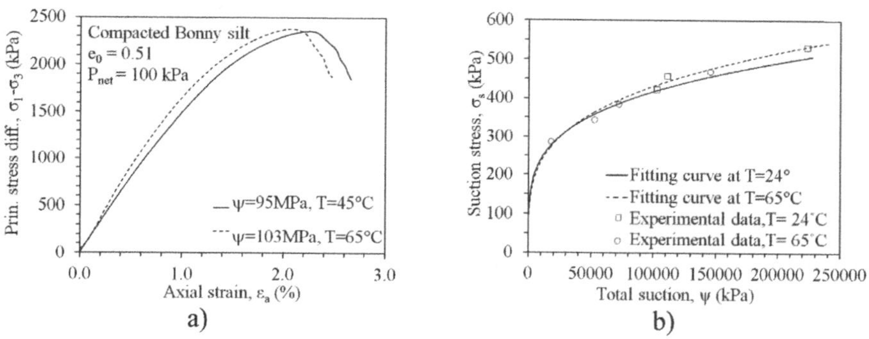

Figure 8.6. *Shear strength of compacted silt under high suction and high temperature. a) Stress–strain curves (e_0 = initial void ratio, P_{net} = net normal stress, ψ = suction, T = temperature). b) Suction–stress characteristic curves*

8.5. Thermal effects on hydraulic properties of unsaturated soils

A linkage between the effective stress in unsaturated soils and the soil-water retention curve (SWRC) has been observed in many studies [VAN 96, NUT 08, KHA 98]. Khalili and Khabbaz [KHA 98] found that the role of suction in the effective stress is strongly dependent on the ratio of the suction to the air entry suction. The air entry suction is an important point on the SWRC. Lu *et al.* [LU 10] derived an equation that integrates the SWRC directly into the effective stress:

$$\sigma' = (\sigma - u_a) + (u_a - u_w) \Big/ \left(1 + \alpha (u_a - u_w)^n\right)^{(n-1)/n} \qquad [8.1]$$

where σ′ and σ are the effective and total stresses, u_a and u_w are the pore air and water pressures, and n and α are the parameters of the van Genuchten [VAN 80] SWRC model. Equation [8.1] can be used to interpret the shear strength of both unsaturated and saturated soils, as this equation reduces to the classic definition of effective stress for saturated conditions. The impact of temperature on the effective stress in unsaturated soils can be evaluated through the SWRC, as the water retention of unsaturated soils decreases with temperature [NIM 86, HOP 86,

GRA 96, ROM 01, BÀC 02, SAL 07, UCH 09]. Specifically, the interface tension and soil–fluid contact angle, variables important in the Kelvin model for suction, are temperature dependent. The interface tension of pure water decreases at a rate of 0.2%/°C [GRA 96], and the fluid–soil contact angle can decrease by as much as 0.26°/°C [BAC 02]. SWRCs for soils defined under constant temperature are shown in Figure 8.7(a). Temperature has a greater impact on the suction magnitude with decreasing degrees of saturation [GRA 96]. The reduction in suction of 100 kPa observed for the compacted silt at a degree of saturation of 0.4 is substantial, indicating that temperature may affect the effective stress. Grant and Salehzadeh [GRA 96] and Salager *et al.* [SAL 07] developed incremental-form models for temperature effects on the SWRC that use Kelvin's equation for suction and consider the relative thermal expansion of soil and water, but do not incorporate common SWRC equations. An area of continued research is the impact of temperature on the SWRC at high suction magnitudes. Alsherif and McCartney [ALS 13] developed a new thermal triaxial cell with suction control using the vapor circulation technique [LIK 03]. The constant relative humidity (Rh) for different temperatures (T) in Figure 8.7(b) reflects the feasibility of maintaining constant high suctions during elevated temperature with this device.

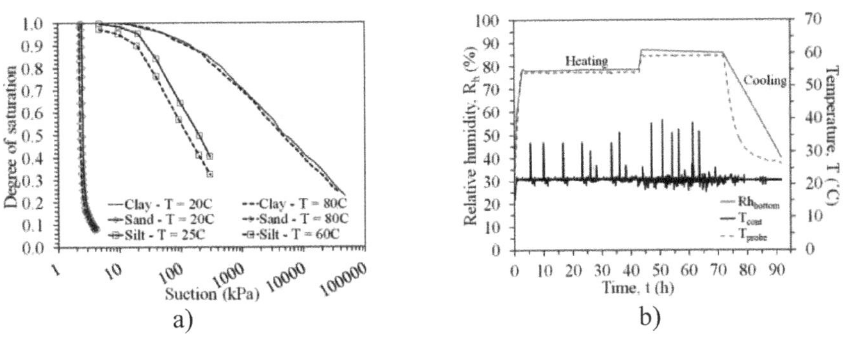

Figure 8.7. *a) Thermal effects on SWRCs of clay [ROM 03], silt [UCH 09] and sand [EWE 89]. b) Thermal effects at high suctions on compacted silt [ALS 12]*

8.6. Thermal effects on soil–geosynthetic interaction

Woven geotextiles or geogrids and non-woven geotextiles are primarily used for reinforcement of poorly draining backfills because extruded geogrids do not interact well with fine-grained soils. Recent studies on reinforcement of poorly draining backfills have focused on measurement of the effects of suction on soil–geosynthetic interface shear strength [HAT 08, SHA 98, HAM 09, KHO 10]. These studies used a two stress-state variable approach to interpret the effects of suction on shear strength (Figures 8.8(a) and (b)). However, the results from the previous section indicate that the impact of temperature may be easier to interpret when the shear strength is interpreted using effective stress. The SWRC of this soil (Figure 8.8(c)) indicates that the interface shear strength can be interpreted well using a single-value effective stress (Figure 8.8(d)). This confirms that the shear strength of unsaturated soil–geosynthetic interfaces is strongly dependent on the matric suction in the soil.

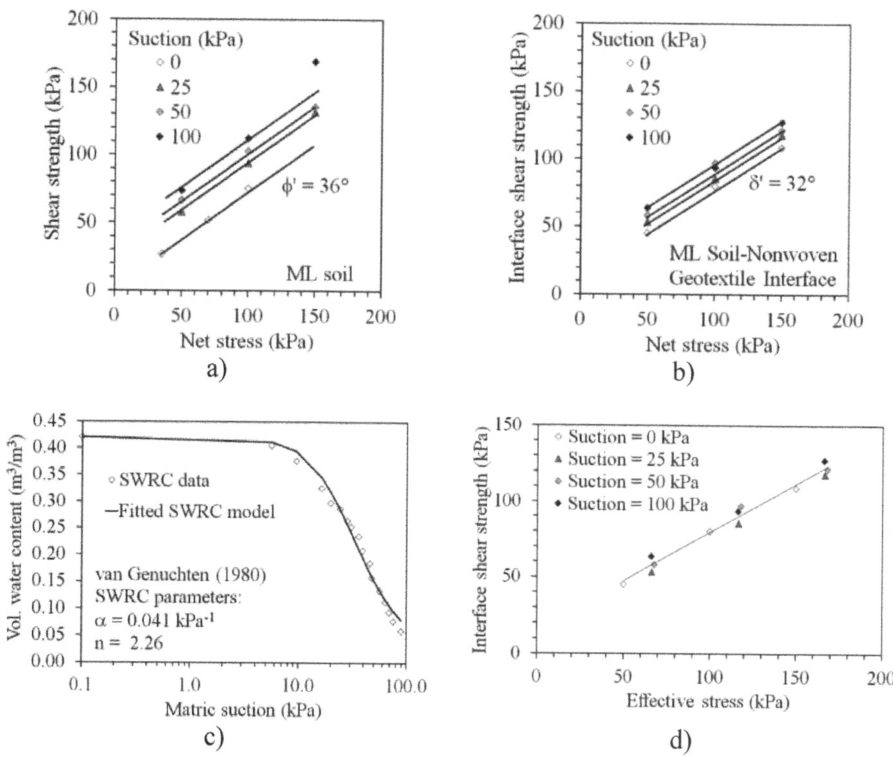

Figure 8.8. *Unsaturated interface shear strength of soil–geosynthetic interfaces [KHO 10]. a) Impact of suction on soil (ϕ' is soil friction angle), b) impact of suction on soil–geosynthetic interface (δ' is soil–geosynthetic interface friction angle), c) SWRC and d) interface shear strength using effective stress*

An added complexity in thermally active geotechnical systems is that the tensile stress–strain behavior of geosynthetics is sensitive to temperature, a feature that has been used to accelerate the creep process using the stepped isothermal method [THO 98, ZOR 04, BUE 05]. The results from unconfined thermal creep tests in Figure 8.9(a) indicate a linear increase in creep strain with temperature, which reflects the potential impacts of temperature on the stress–strain behavior. Despite this observed behavior, confinement of the geosynthetic in soil may reduce the impact of thermally induced creep by preventing geometric distortion of the geosynthetic structure [MCG 82]. The data in Figure 8.9(b) show that confinement not only leads to a lower initial tensile strain during application of a relatively high fraction of the ultimate tensile strength P_{ult}, but also to a lower rate of creep strain over time [FRA 11].

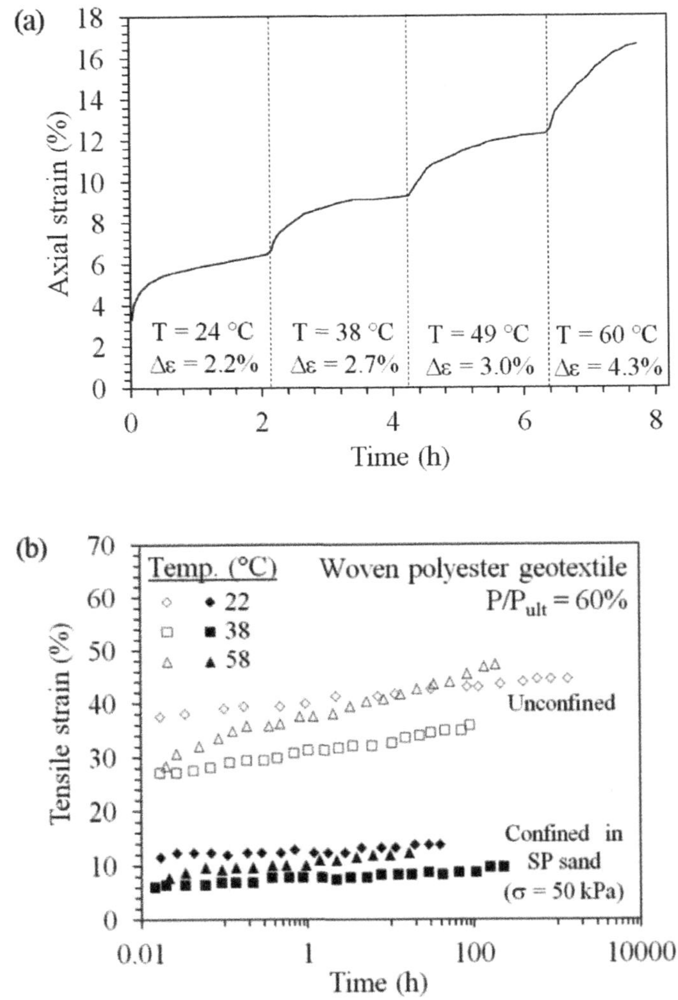

Figure 8.9. *a) Thermally induced creep strain (Δε) of a geotextile [ZOR 04]. b) Impact of confining stress on creep of a non-woven geotextile (σ = confining stress, P = tensile stress, P_{ult} = ultimate tensile strength) [FRA 11]*

8.7. Conclusions

Some of the key issues involved in the development of thermally active geotechnical systems in unsaturated soils are summarized in this chapter. Additional research is required to synthesize effects of the different thermohydromechanical processes in unsaturated, reinforced soils, and to understand the interaction between unsaturated conditions and thermally induced creep deformations of confined geosynthetics. Nonetheless, the results from the literature indicate that the strategy to maintain unsaturated conditions in soil systems using thermally induced water flow will enable, under the right conditions, use of a broader class of backfills than currently permitted in fill-type geotechnical applications. Furthermore, although temperature may have some negative impacts from the perspective of thermal soil improvement, the positive effects of increased suction due to thermally induced water flow may be sufficient to compensate for these effects. An added benefit of thermally active geotechnical systems is that they can dissipate spurious heat generated by buildings or industry.

8.8. Acknowledgments

The authors appreciate the support of the National Science Foundation grant CMMI-1054190. The views in this chapter are those of the authors alone.

8.9. Bibliography

[ABD 81] ABDEL-HADI O., MITCHELL J. "Coupled heat and water flows around buried cables", *Journal of the Soil Mechanics and Foundations Engineering Division*, vol. 107, no. 11, pp. 1461–1487, 1981.

[ABU 07a] ABUEL-NAGA H., BERGADO D., BOUAZZA A., *et al.*, "Volume change behavior of saturated clays under drained heating conditions: experimental results and constitutive modeling", *Canadian Geotechnical Journal*, vol. 44, no. 8, pp. 942–956, 2007.

[ABU 07b] ABUEL-NAGA H., BERGADO D., LIM B. "Effect of temperature on shear strength and yielding behavior of soft Bangkok clay", *Soils and Foundations*, vol. 47, no. 3, pp. 423–436, 2007.

[ABU 09] ABUEL-NAGA H., BERGADO D., BOUAZZA A., *et al.*, "Thermo-mechanical model for saturated clays", *Géotechnique*, vol. 59, no. 3, pp. 273–278, 2009.

[ALO 90] ALONSO E., GENS A., JOSA A., "A constitutive model for partially saturated soils", *Géotechnique*, vol. 40, pp. 405–430, 1990.

[ALS 12] ALSHERIF N., MCCARTNEY J. "Nonisothermal shear strength of soils under high suctions", *2nd European Unsaturated Soils Conference*, Napoli, p. 8, 21–22 June 2012.

[ALS 13] ALSHERIF N., MCCARTNEY J., "Triaxial cell for nonisothermal shear strength of compacted silt under high suction magnitudes", *1st Pan-American Conference on Unsaturated Soils*, Cartagena de Indias, Colombia, 20–22 February 2013.

[BAC 02] BACHMANN J., HORTON R., GRANT S., *et al.*, "Temperature dependence of water retention curves for wettable and water-repellent soils", *Soil Science Society of America Journal*, vol. 66, pp. 44–52, 2002.

[BAL 88] BALDI G., HUECKEL T., PELLEGRINI R., "Thermal volume changes of the mineral-water system in low-porosity clay soils", *Canadian Geotechnical Journal*, vol. 25, pp. 807–825, 1988.

[BUE 05] BUENO B., COSTANZI M., ZORNBERG J., "Conventional and accelerated creep tests on nonwoven needlepunched geotextiles", *Geosynthetics International*, vol. 12, no. 6, pp. 276–287, 2005.

[CAM 68] CAMPANELLA R., MITCHELL J., "Influence of temperature variations on soil behavior", *Journal of the Soil Mechanics and Foundations Engineering Division*, vol. 94(3), pp. 709–734, 1968.

[CAR 62a] CARY J., TAYLOR S., "The interaction of the simultaneous diffusions of heat and water vapor", *Soil Science Society of America Proceedings*, vol. 26, pp. 413–416, 1962.

[CAR 62b] CARY J., TAYLOR S., "Thermally driven liquid and vapor phase transfer of water and energy in soil", *Soil Science Society of America Proceedings*, vol. 26, pp. 417–420, 1962.

[CEK 04] CEKEREVAC C., LALOUI L., "Experimental study of thermal effects on the mechanical behavior of a clay", *International Journal for Numerical Analytical Methods in Geomechanics*, vol. 28, pp. 209–228, 2004.

[COC 13] COCCIA C., MCCARTNEY J., "Impact of heat exchange on the thermo-hydromechanical response of reinforced embankments", *GeoCongress 2013*, ASCE, San Diego, CA, p. 10, 3–5 March 2013.

[CUI 00] CUI Y., SULTAN N., DELAGE P., "A thermomechanical model for clays", *Canadian Geotechnical Journal*, vol. 37, no. 3, pp. 607–620, 2000.

[CUI 09] CUI Y., LE T., TANG A., *et al.*, "Investigating the time-dependent behaviour of Boom clay under thermo-mechanical loading", *Géotechnique*, vol. 59, no. 4, pp. 319–329, 2009.

[DEL 00] DELAGE P., SULTAN N., CUI Y., "On the thermal consolidation of Boom clay", *Canadian Geotechnical Journal*, vol. 37, pp. 343–354, 2000.

[DEM 82] DEMARS K., CHARLES R., "Soil volume changes induced by temperature cycling", *Canadian Geotechnical Journal*, vol. 19, pp. 188–194, 1982.

[DRB 96] DRBAL L., WESTRA K., BOSTON P., *Power Plant Engineering*, 1st ed., Springer, New York, 1996.

[ELI 01] ELIAS V., CHRISTOPHER B., BERG R., Mechanically stabilized earth walls and reinforced soil slopes: design and construction guidelines, Report No. FHWA-NHI-00-043, Federal Highway Administration, Washington DC, p. 418, 2001.

[ERI 89] ERIKSSON L., "Temperature effects on consolidation properties of sulphide clays", *Proceedings of the 12th International Conference on Soil Mechanics and Foundation Engineering*, vol. 3, Rio de Janeiro, Taylor and Francis , pp. 2087–2090, 1989.

[EWE 89] EWEN J., THOMAS H., "Heating unsaturated medium sand", *Géotechnique*, vol. 39, no. 3, pp. 455–470, 1989.

[FAR 81] FAROUKI O., *Thermal Properties of Soils*, US Army Corps of Engineers, CRREL Monograph 81-1, 1981.

[FIN 51] FINN F., "The effects of temperature on the consolidation characteristics of remolded clay", *Symposium on Consolidation Testing of Soils*, ASTM STP 126, pp. 65–72, 1951.

[FRA 05] FRANCOIS B., SALAGER S., EL YOUSSOUFI M.S., *et al.*, "Compression tests on a sandy silt at different suction and temperature levels", *Proc. GeoDevers 2007, GSP 157: Computer Applications in Geotechnical Engineering*, ASCE, p. 10, 2005.

[FRA 08] FRANCOIS B., LALOUI L., "Unsaturated soils under non-isothermal conditions: framework of a new constitutive model", *GeoCongress 2008*, ASCE, New Orleans, LA, pp. 1077–1083, 2008.

[FRA 11] FRANCA F., BUENO B., ZORNBERG J., "Journal confinadoes e aceleradoes de fluencia em geossinteticos", *Revista Fundações e Obras Geotécnicas*, vol. 2, no. 12, pp. 56–63, 2011.

[GRA 96] GRANT S., SALEHZADEH A., "Calculations of temperature effects on wetting coefficients of porous solids and their capillary pressure functions", *Water Resources Research*, vol. 32, pp. 261–279, 1996.

[GRA 01] GRAHAM J., TANAKA N., CRILLY T., *et al.*, "Modified Cam-Clay modeling of temperature effects in clays", *Canadian Geotechnical Journal*, vol. 38, no. 3, pp. 608–621, 2001.

[HAM 09] HAMID T., MILLER G., "Shear strength of un-saturated soil interfaces", *Canadian Geotechnical Journal*, vol. 46, pp. 595–606, 2009.

[HAT 08] HATAMI K., KHOURY C., MILLER G., "Suction-controlled testing of soil-geotextile interfaces", *GeoAmericas 2008*, Cancun, Mexico, p. 10, 2008.

[HOP 86] HOPMANS J., DANE J., "Temperature dependence of soil hydraulic properties", *SSSA Journal*, vol. 50, pp. 4–9, 1986.

[HOU 85] HOUSTON S., HOUSTON W., WILLIAMS N., "Thermo-mechanical behavior of seafloor sediments", *Journal of Geotechnical Engineering*, vol. 111, no. 12, pp. 1249–1263, 1985.

[HUE 90a] HUECKEL T., BALDI M., "Thermoplasticity of saturated clays: experimental constitutive study", *Journal of Geotechnical Engineering*, vol. 116, no. 12, pp. 1778–1796, 1990.

[HUE 90b] HUECKEL T., BORSETTO M., "Thermoplasticity of saturated soils and shales: constitutive equations", *Journal of Geotechnical Engineering*, vol. 116, no. 12, pp. 1765–1777, 1990.

[HUE 92] HUECKEL T., PELLEGRINI R., "Effective stress and water pressure in saturated clays during heating-cooling cycles", *Canadian Geotechnical Journal*, vol. 29, pp. 1095–1102, 1992.

[HUE 09] HUECKEL T., FRANÇOIS B., LALOUI L., "Explaining thermal failure in saturated clays", *Géotechnique*, vol. 59, no. 3, pp. 197–212, 2009.

[KHA 98] KHALILI N., KHABBAZ M.H., "A unique relationship for the determination of shear strength of unsaturated soils", *Géotechnique*, vol. 48, pp. 681–688, 1998.

[KHA 04] KHALILI N., GEISER F., BLIGHT G., "Effective stress in unsaturated soils: a review with new evidence", *International Journal of Geomechanics*, vol. 4, no. 2, pp. 115–126, 2004.

[KHO 10] KHOURY C., MILLER G., HATAMI K., "Shear strength of unsaturated soil-geotextile interfaces", *GeoFlorida 2010, Advances in Analysis, Modeling and Design, GSP 199*, pp. 307–316, 2010.

[KUN 95] KUNTIWATTANAKUL P., TOWHATA I., OHISHI K., *et al.*, "Temperature effects on undrained shear characteristics of clay", *Soils and Found*, vol. 35, no. 1, pp. 147–162, 1995.

[LAL 03] LALOUI L., CEKEREVAC C., "Thermo-plasticity of clays: an isotropic yield mechanism", *Computers and Geotechnics*, vol. 30, no. 8, pp. 649–660, 2003.

[LIK 03] LIKOS W., LU N., "Automated humidity system for measuring total suction characteristics of clay", *Geotechnical Testing Journal*, vol. 26, no. 2, pp. 1–12, 2003.

[LU 06] LU N., LIKOS W., "Suction stress characteristic curve for unsaturated soil", *Journal of Geotechnical and Geoenvironmental Engineering*, vol. 132, no. 2, pp. 131–142, 2006.

[LU 10] LU N., GODT J., WU D., "A closed-form equation for effective stress in unsaturated soil", *Water Resources Research*, vol. 46, p. 14, 2010.

[MCC 04] MCCARTNEY J., ZORNBERG J., "Use of moisture profiles and lysimetry to assess evapotranspirative cover performance", *5th International PhD Symposium in Civil Engineering*, Delft, the Netherlands, pp. 961–969, 2004.

[MCG 82] MCGOWN A., ANDRAWES K., KABIR M., "Load-extension testing of geotextiles confined in soil", *2nd International Conference on Geotextiles*, Las Vegas, NV, pp. 793–798, 1982.

[MCK 65] MCKINSTRY H., "Thermal expansion of clay minerals", *The American Mineralogist*, vol. 50, pp. 212–222, 1965.

[MIL 82] MILLY P., "Moisture and heat transport in hysteretic, inhomogeneous porous media: A matric head-based formulation and a numerical model", *Water Resources Research*, vol. 18, no. 3, pp. 489–198, 1982.

[MIL 96] MILLY P., "Effects of thermal vapor diffusion on seasonal dynamics of water in the unsaturated zone", *Water Resources Research*, vol. 32, no. 3, pp. 509–518, 1996.

[MIL 07] MILLER G., HAMID T., "Interface direct shear testing of unsaturated soil", *Geotechnical Testing Journal*, vol. 30, no. 3, pp. 1–10, 2007.

[MIT 95] MITCHELL J., ZORNBERG J., "Reinforced soil structures with poorly draining backfills. Part II: case histories and applications", *Geosynthetics International*, vol. 2, no. 1, pp. 265–307, 1995.

[MIT 05] MITCHELL J., SOGA K., *Fundamentals of Soil Behavior*, 3rd ed., John Wiley & Sons, Inc., New York, 2005.

[NIM 86] NIMMO J., MILLER E., "The temperature dependence of isothermal moisture vs. potential characteristics of soils", *SSSA Journal*, vol. 50, pp. 1105–1113, 1986.

[NRE 97] NRECA, Geothermal Heat Pumps: Introductory Guide, RER Project 86-1A, National Rural Electric and Cooperative Association, p. 99, 2007.

[NUT 08] NUTH M., LALOUI L., "Effective stress concept in unsaturated soils: clarification and validation of a unified framework", *International Journal for Numerical and Analytical Methods in Geomechanics*, vol. 32, pp. 771–801, 2008.

[OGO 08] OGORZALEK A., BOHNHOFF G., SHACKELFORD C., *et al.*, "Comparison of field data and water-balance predictions for a capillary barrier cover", *Journal of Geotechnical and Geoenvironmental Engineering*, vol. 134, no. 4, pp. 470–486, 2008.

[OME 08] OMER A., "Ground-source heat pumps systems and applications", *Renewable and Sustainable Energy Reviews*, vol. 12, no. 2, pp. 344–371, 2008.

[PAA 67] PAASWELL R., "Temperature effects on clay soil consolidation", *Journal of the Soil Mechanics and Foundation Engineering Division*, vol. 93, no. 3, pp. 9–22, 1967.

[PHI 57] PHILIP J., DE VRIES D., "Moisture movement in porous materials under temperature gradients", *Transactions of the American Geophysical Union*, vol. 38, no. 2, pp. 222–232, 1957.

[PLU 69] PLUM R., ESRIG M., Some temperature effects on soil compressibility and pore water pressure, Report 103, Highway Research Board, Washington, DC, pp. 231–242, 1969.

[ROM 01] ROMERO E., GENS A., LLORET A., "Temperature effects on the hydraulic behavior of an unsaturated clay", *Geotechnical and Geological Engineering*, vol. 19, pp. 311–332, 2001.

[ROM 03] ROMERO E., GENS A., LLORET A., "Suction effects on a compacted clay under non-isothermal conditions", *Géotechnique*, vol. 53, no. 1, pp. 65–81, 2003.

[ROM 05] ROMERO E., VILLAR M., LLORET A., "Thermo-hydro-mechanical behavior of heavily overconsolidated clays", *Engineering Geology*, vol. 81, pp. 255–268, 2005.

[SAB 97] SABATINI P., ELIAS V., SCHMERTMANN G., *et al.*, *Geotechnical Engineering Circular Number 2: Earth Retaining Systems*, FHWA, Washington, DC, 1997.

[SAI 90] SAIX C., JOUANNA P., "Appareil triaxial pour l'étude du comportement thermique de sols non saturés", *Canadian Geotechnical Journal*, vol. 27, pp. 119–128, 1990.

[SAI 91] SAIX C., "Consolidation thermique par chaleur d'un sol non saturé", *Canadian Geotechnical Journal*, vol. 28, pp. 42–50, 1991.

[SAI 00] SAIX C., DEVILLERS P., EL YOUSSOUFI M., "Élément de couplage thermomécanique dans la consolidation de sols non saturés", *Canadian Geotechnical Journal*, vol. 37, pp. 308–317, 2000.

[SAL 07] SALAGER S., EL YOUSSOUFI M., SAIX C., "Influence of temperature on the water retention curve of soils: modelling and experiments", SCHANZ T. (ed.), *Experimental Unsaturated Soil Mechanics*, Springer, pp. 251–258, 2007.

[SCA 02] SCANLON B., CHRISTMAN M., REEDY R., *et al.*, "Intercode comparisons for simulating water balance of surficial sediments in semiarid regions", *Water Resources Research*, vol. 38, no. 12, pp. 1–16, 2002.

[SHA 07] SHARMA J., FLEMING I., JOGI M., "Measurement of unsaturated soil-geomembrane interface shear strength parameters", *Canadian Geotechnical Journal*, vol. 44, pp. 78–88, 2007.

[SMI 13] SMITS K., SAKAKI T., HOWINGTON S., *et al.*, "Temperature dependence of thermal properties of sands across a wide range of temperatures (30–70°C). *Vadose Zone Journal*, vol. 12, no. 1, p. 8, 2013.

[SUL 02] SULTAN N., DELAGE P., CUI Y., "Temperature effects on the volume change behavior of Boom clay", *Engineering Geology*, vol. 64, pp. 135–145, 2002.

[TAN 08] TANG A., CUI Y., BARNEL N., "Thermo-mechanical behavior of a compacted swelling clay", *Géotechnique*, vol. 58, no. 1, pp. 45–54, 2008.

[TAN 09] TANG A., CUI Y., "Modelling the thermomechanical volume change behaviour of compacted expansive clays", *Géotechnique*, vol. 59, no. 3, pp. 185–195, 2009.

[TAY 64] TAYLOR S., CARY J., "Linear equations for the simultaneous flow of water and energy in continuous soil system", *Soil Science Society of America Proceedings*, vol. 28, pp. 167–172, 1964.

[THO 95a] THOMAS H., HE Y., "Analysis of coupled heat, moisture, and air transfer in a deformable unsaturated soil", *Géotechnique*, vol. 45, no. 4, pp. 677–689, 1995.

[THO 95b] THOMAS H., SANSOM M., "Fully coupled heat, moisture, and air transfer in an unsaturated soil", *Journal of Engineering Mechanics*, vol. 12, no. 3, pp. 392–405, 1995.

[THO 96] THOMAS H., HE Y., SANSOM M., *et al.*, "On the development of a model of the thermo-mechanical-hydraulic behavior of unsaturated soils", *Engineering Geology*, vol. 41, pp. 197–218, 1996.

[THO 98] THORNTON J., ALLEN S., THOMAS R., *et al.*, "The stepped isothermal method for time-temperature superposition and its application to creep data on polyester yarn", *6th International Conference on Geosynthetics*, IGS, Atlanta, pp. 699–706, 1998.

[TID 89] TIDFORS M., SÄLLFORS G., "Temperature effect on preconsolidation pressure", *Geotechnical Testing Journal*, vol. 12, no. 1, pp. 93–97, 1989.

[TOW 93] TOWHATA I., KUNTIWATTANAKUL P., SEKO I., *et al.*, "Volume change of clays induced by heating as observed in consolidation tests", *Soils and Foundations*, vol. 33, no. 4, pp. 170–183, 1993.

[UCH 09] UCHAIPICHAT A., KHALILI N., "Experimental investigation of thermo-hydro-mechanical behaviour of an unsaturated silt", *Géotechnique*, vol. 59, no. 4, pp. 339–353, 2009.

[VAN 96] VANAPALLI S., FREDLUND D., PUFAHL D., *et al.*, "Model for the prediction of shear strength with respect to soil suction", *Canadian Geotechnical Journal*, vol. 33, pp. 379–392, 1996.

[VAN 80] VAN GENUCHTEN M., "A closed-form equation for predicting the hydraulic conductivity of unsaturated soils", *Soil Science Society of America Journal*, vol. 44, pp. 892–898, 1980.

[WIL 94] WILSON G., FREDLUND D., BARBOUR S., "Coupled soil-atmosphere modeling for soil evaporation", *Canadian Geotechnical Journal*, vol. 31, pp. 151–161, 1994.

[YON 96] YONG R., MOHAMED A., "Evaluation of coupled heat and moisture flow patterns in a bentonite-sand buffer material", *Engineering Geology*, vol. 41, pp. 269–286, 1996.

[ZOR 94] ZORNBERG J., MITCHELL J., "Reinforced soil structures with poorly draining backfills. Part I: reinforcement interactions and functions", *Geosynthetics International*, vol. 1, no. 2, pp. 103–148, 1994.

[ZOR 95] ZORNBERG J., CHRISTOPHER B., MITCHELL J., "Performance of a geotextile-reinforced slope using de-composed granite as backfill material", *2nd Brazilian Symposium on Geosynthetics*, São Paulo, Brazil, pp. 19–29, July 1995.

[ZOR 04] ZORNBERG J., BYLER B., KNUDSEN J., "Creep of geotextiles using time-temperature superposition methods", *Journal of Geotechnical and Geoenvironmental Engineering*, vol. 130, no. 11, pp. 1158–1168, 2004.

[ZOR 05] ZORNBERG J., MCCARTNEY J., "Evaluation of evapotranspiration from alternative landfill covers at the Rocky Mountain Arsenal", *Experus 2005, Trento, Italy*, Balkema, Rotterdam, 27–29 June 2005.

Chapter 9

Energy Geostructures in Cooling-Dominated Climates

9.1. Introduction

The use of energy geostructures as ground heat exchangers has achieved success in Europe over the last two decades, where the primary application is for heating. Developing and designing these systems in other parts of the world requires considering regional climatic conditions. In cold climates, such as northern Europe, the heating requirement is more significant than the need for cooling; the reverse is true in hot regions. The application of energy geostructures for heating or cooling is equivalent from the thermodynamic perspective, but differences arise in implementation. In cooling-dominated climates, high temperature, high evapotranspiration, sporadic rainfall and water table fluctuation affects the soil conditions, and therefore the efficiency of energy geostructures as heat exchangers. In addition, the imbalanced heat injection/extraction in the ground results in an increase in soil temperature over the years of system operation, which affects the long-term performance of energy geostructures. This chapter presents an overview of the impact of climatic factors on soil conditions, and the associated energy geostructures' performance in different soil types and situations.

9.2. Climatic factors and their effects on soil conditions and properties

Different climatic zones exist across the world, and they differ in the following four main aspects: (1) air temperature, (2) rainfall, (3) evapotranspiration and

---

Chapter written by Ghassan Anis AKROUCH, Marcelo SANCHEZ and Jean-Louis BRIAUD.

(4) relative humidity (RH). Each of these four factors affects thermal properties, moisture content (degree of saturation profile) and the water table in the shallow part of the soil, referred to from here on as the shallow soil layer (SSL). For traditional long boreholes used as heat exchangers, the soil can be assumed to be homogeneous during the design process, because climate-induced variations occur at a shallow depth relative to the total length of the borehole. This assumption is not valid for energy geostructures. The mechanical and thermodynamic performance of energy geostructures is significantly affected by soil variability, since a greater proportion of these structures is in direct contact with the SSL.

Humid or dry, hot climates are characterized by high yearly average temperature. In any climate, below a certain depth soil temperature is moderate, constant and equal to the annual average air temperature. This depth is referred to here as the thermal independence (TI) depth. The location of the TI depth is a factor of soil type and thermal properties. Geothermal energy systems in cooling-dominated climates work by circulating hot fluid in pipes, and relying on a temperature gradient to inject heat into the soil. Therefore, higher soil temperature results in lower system efficiency for cooling.

Rainfall distribution, and the corresponding elevation and fluctuation of the ground water table, is another parameter that impacts soil properties and conditions. Well-distributed rainfall and snow, characteristic of cold climates, keeps the SSL saturated over the year. Conversely, rainfalls in hot climates are infrequent and sporadic, which significantly impact its saturation profile.

Furthermore, evapotranspiration, which is the sum of evaporation and transpiration, impacts the soil moisture profile. Due to the high temperature in hot climates, water evaporates and trees and grass pull water from the ground, which results in soil drying. This influences the moisture content profile in the SSL. Changes in RH also affect soil moisture. Soil suction is related to RH through the psychometric law, and the water content (or degree of saturation) depends on suction via the soil water retention curve [FRE 93, LU 04]

In hot climates, the thermal load on an energy geostructure is unbalanced, because the demand for cooling is dominant. Greater heat injection than extraction results in a progressive increase in soil and pile temperatures. The increase in soil temperature causes a reduction in the long-term thermodynamic performance of the system, due to the decreasing heat exchange rate between energy geostructures and the soil. The increase in pile temperature induces additional stresses and strains that can result in unwanted deformation of the upper structure. To overcome this problem, energy geostructures can be designed for lower heat exchange, or use hybrid systems.

In conclusion, climatic factors impact soil conditions in the SSL, especially the soil saturation profile (Figure 9.1). Due to the direct relationship between soil thermal properties and degree of saturation, these factors should be taken into account when designing energy geostructures, especially in hot climates.

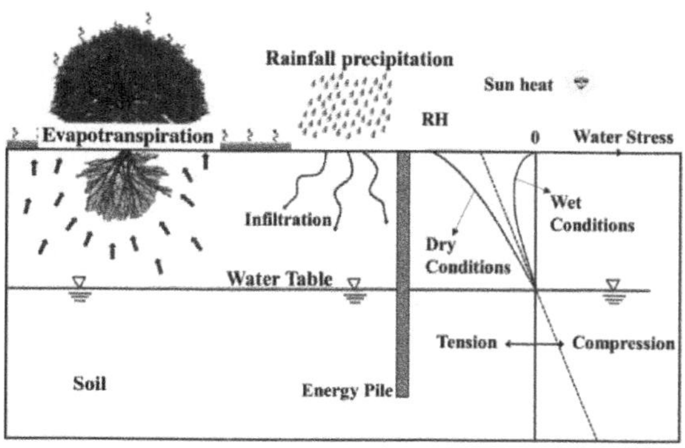

Figure 9.1. *Climatic factors and effect on soil conditions*

9.3. Saturated and unsaturated soil thermal properties and heat transfer

Soil is a porous multiphase material constituted of three main phases: solid (soil grains), liquid (water) and gas (air). When soil is fully saturated, all pores are filled with water; when soil is fully dry, all pores are filled with air; and when soil is partially saturated, pores are filled with a mixture of water and air. Heat transfer in soils can take place by conduction, convection and radiation. Radiation heat transfer is negligible in soils, and its effect in sands is less than 1% of the overall heat transfer [REE 00]. However, convection heat transfer is significant when groundwater flow conditions exist. Heat conduction is the more relevant process associated with heat transfer in soils. It is controlled by the well-known Fourier's law [FOU 22]. Knowing that thermal properties and their relation to temperature are different for each phase that constitutes the soil, the effective soil thermal properties are dependent on the volumetric fraction of each soil phase. The relationship between soil thermal conductivity and water content can be explained by solid particle and water interaction (Figure 9.2). At a very low water content (the exact value of which depends on soil texture and soil-specific surface area referred to here as "θ_1"), the water film thickness is very thin and not enough to improve the contact between soil particles. Therefore, the soil thermal conductivity remains constant up to θ_1. After this water content threshold is exceeded, water bridges between solid particles start developing and increasing, which results in a rapid increase in thermal conductivity. Eventually, the increase in water content depends on the displacement of air by water. When this happens, the rate of increase in thermal conductivity slows [SEP 79, TAR 02].

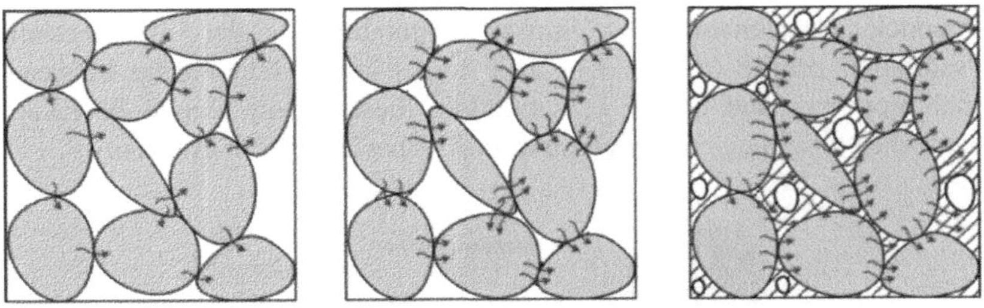

Figure 9.2. *Heat flow in soil under different conditions*

Different models have been proposed to predict thermal properties of unsaturated soils. For example, an empirical model proposed by Côté and Konrad [CÔT 05] (equations [9.1] and [9.2]) relates the effective thermal conductivity, λ_e, to soil degree of saturation S_l, soil porosity ϕ, soil texture (represented by κ) and soil particle shape parameters (represented by χ and η). Note that Johansen [JOH 75] was the first to introduce the effective thermal conductivity concept. The values of κ, χ and η for different soils are presented in Table 9.1. In equation [9.1], λ, λ_{sat}, λ_{dry} and λ_s are soil, saturated, dry and solids thermal conductivity, respectively.

$$\lambda_e = \frac{\lambda - \lambda_{dry}}{\lambda_{sat} - \lambda_{dry}} = f(S_l), \text{ with } f(S_l) = \frac{\kappa S_l}{1 + (\kappa - 1) S_l} \quad [9.1]$$

$$\lambda_{sat} = \lambda_s^{1-\phi} \times 0.6^{\phi} \text{ and } \lambda_{dry} = \chi 10^{-\eta\phi} \quad [9.2]$$

Soil type	κ		χ	η
	Unfrozen	Frozen		
Gravels and coarse sands	4.6	1.70	1.70	1.80
Medium and fine sands	3.55	0.95	0.75	1.20
Silty and clayey soils	1.90	0.85	0.75	1.20
Organic fibrous soils (peat)	0.60	0.25	0.30	0.87

Table 9.1. *Values of soil texture and shape parameters (after [CÔT 05])*

9.4. Impact of soil conditions on energy geostructures performance

Energy geostructures are sized and designed to satisfy the building structural and thermal load. A change in soil conditions will impact the structural capacity and overall thermodynamic performance of energy piles. The latter is the focus of the remainder of this chapter. To evaluate the impact of soil conditions on energy geostructures performance, a series of laboratory tests and numerical simulation were conducted at Texas A&M University. To study the behavior of an energy pile at a certain depth and a given soil condition, laboratory tests were performed on a soil-energy pile planar section in sand. The test was performed for different scenarios based on varying the degree of soil saturation, S_l, to represent conditions ranging from fully dry to fully saturated. The laboratory tests were then numerically modeled to evaluate the heat exchange rate between the soil and energy pile. The results from the planar section laboratory test and numerical model were used to develop the relationship between the heat transfer rate and soil degree of saturation. The numerical model was extended to account for the soil and pile vertical profile. From the numerical results, the heat flux and saturation profile were evaluated. In addition, the energy pile efficiency ratio, ε, was evaluated. This ratio is defined as the ratio between the heat exchanged when the soil is partially saturated, Q_{unsat}, and when the soil is saturated, Q_{sat}.

9.4.1. *Laboratory experimental design*

The laboratory test consisted of an energy pile concrete section poured in the lab and put in a wooden box filled with sand. The energy pile section was fitted with two PVC pipes simulating energy pile. The wood box was covered from the top and the bottom by two insulation boards to reduce the heat transfer to a two-dimensional (2D) problem. The PVC pipes were filled with water and instrumented by an aquarium heater with regulators to control the water temperature. The water in the pipes was heated to a constant temperature of 37°C, and the test was performed for 48 h and repeated for 12 different soil saturation conditions. The temperature at the pile–soil interface and in the soil at different locations was recorded using a datalogger (points B, C, D, E, F and G). The laboratory test was performed on different pile sections and different soil conditions ranging from fully dry to fully saturated conditions. This chapter will present the results for a 0.3 m diameter section pile and three soil saturation conditions: fully dry soil (Test 1), partially saturated soil (Test 4) and fully saturated soil (Test 6).

The soil material used in the laboratory test was medium fine sand with engineering and thermal properties presented in Table 9.2. The soil thermal conductivity was evaluated using the Shannon and Wells [SHA 47] method. For this method, a cylindrical soil sample was put in a constant-temperature oven, and the increase at the center of the sample was recorded. Then, the thermal properties were back-calculated based on the analytical solution of this problem. The concrete and PVC pipes used in the laboratory test had a thermal conductivity of 1.4 and 0.15 W/m·K, respectively.

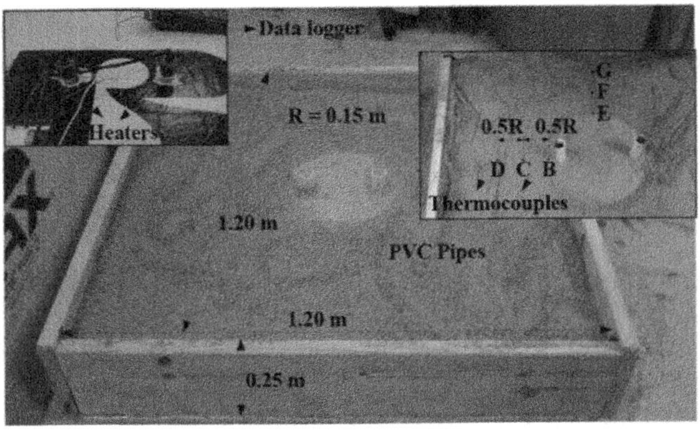

Figure 9.3. *Laboratory test setup*

Test ID	k (W/m·K)	S_l	ϕ	Void ratio, e	Water content ω (%)	Total/dry unit weight, γ_T/γ_d (kN/m^3)
Test 1	0.90	0.01	0.45	0.8	0.5	14.74/14.68
Test 4	2.1	0.48	0.45	0.81	14.5	16.72/14.60
Test 6	2.65	1.00	0.45	0.81	30.0	19.02/14.62

Table 9.2. *Soil engineering and thermal properties for the different test*

9.4.2. *Numerical modeling*

The numerical modeling was performed using the finite element program CODE_BRIGHT that solves coupled thermo-hydro-mechanical (THM) problems in porous media [OLI 96, GEN 98, ALO 99]. In the numerical simulation performed, the mechanical problem was not taken into account. The THM theoretical framework was composed of two main parts: balance equations and constitutive equations. The balance equations of the total water mass and energy were expressed as equations [9.3] and [9.4], respectively. Note that the formulation presented in this chapter is general; more details can be found in [OLI 96].

$$\frac{\partial}{\partial t}\left(\theta_l^w S_l \phi + \theta_g^w S_g \phi\right) + \nabla\left(j_l^w + j_g^w\right) = f^w \qquad [9.3]$$

$$\frac{\partial}{\partial t}\left(E_s \rho_s (1-\phi) + E_l \rho_l S_l \phi + E_g \rho_g S_g \phi\right) + \nabla\left(\boldsymbol{i}_c + \boldsymbol{j}_{Es} + \boldsymbol{j}_{El} + \boldsymbol{j}_{Eg}\right) = f^E \qquad [9.4]$$

where:

θ_l^w	=	mass of water per unit volume of liquid
θ_g^w	=	mass of water per unit volume of gas
j_l^w	=	total mass flux of water in the liquid phase
j_g^w	=	total mass flux of water in the gas phase
E_s, E_l, E_g	=	internal energy of the solid, liquid and gas phase, respectively
i_c	=	conductive heat flow
ϕ	=	porosity
ρ_s, ρ_l, ρ_g	=	solid, liquid and gas density, respectively
j_{Es}, j_{El}, j_{Eg}	=	net flow of energy in mineral, in liquid, and in gas, respectively
f^E	=	the internal/external supply of energy

The constitutive equations used in the Thermo-Hydraulic (TH) model were Darcy's law (equations [9.5] and [9.6]) for liquid flow, and Fourier's law (equation [9.7]) for conductive heat flow.

$$\boldsymbol{q}_l = -\boldsymbol{K}_l\left(\nabla P_l - \rho_l g\right) \qquad [9.5]$$

where $\boldsymbol{q}_l$ is the advective liquid flow, P_l is the liquid pressure, ρ_l is the liquid density, g is the gravity and $\boldsymbol{K}_l$ is the permeability tensor that can be expressed as a function of the intrinsic permeability, k, and the relative permeability function, $k_{r,l}$, that defines the variation of $\boldsymbol{k}$ with liquid saturation condition.

$$\boldsymbol{K}_l = \boldsymbol{k}\frac{k_{r,l}}{\mu_l} \text{ with } k_{r,l} = S_e^{\phi} \text{ where } S_e = \frac{S_l - S_{lr}}{S_{ls} - S_{lr}} \qquad [9.6]$$

$$\boldsymbol{i}_c = -\lambda \nabla T \text{ with } \lambda = \lambda_{sat}\sqrt{S_l} + \lambda_{dry}\left(1 - \sqrt{S_l}\right) \qquad [9.7]$$

where S_e, S_{lr} and S_{ls} are the effective, minimum or residual and maximum degree of saturation, respectively, T is the temperature and μ_l is the water viscosity. Hydrostatic water condition was used in the numerical model; therefore, the matrix suction profile above the ground water table can be expressed as in equation [9.8]. The van Genuchten model (equation [9.9]) was used in the numerical simulation to relate the degree of saturation to suction ψ.

$$\psi = P_g - P_l \qquad [9.8]$$

$$S_e = \left[1 + \left(\frac{\psi}{P_0}\right)^{\frac{1}{1-\lambda_0}}\right]^{-\lambda_0} \qquad [9.9]$$

where P_g, P_0 and λ_0 are the gas pressure, air entry value and van Genuchten model fitting parameter, respectively. The planar section and full pile numerical models with initial and boundary conditions are shown in Figure 9.4. The planar section model was a plain strain model, while the full pile model was an axisymmetric model.

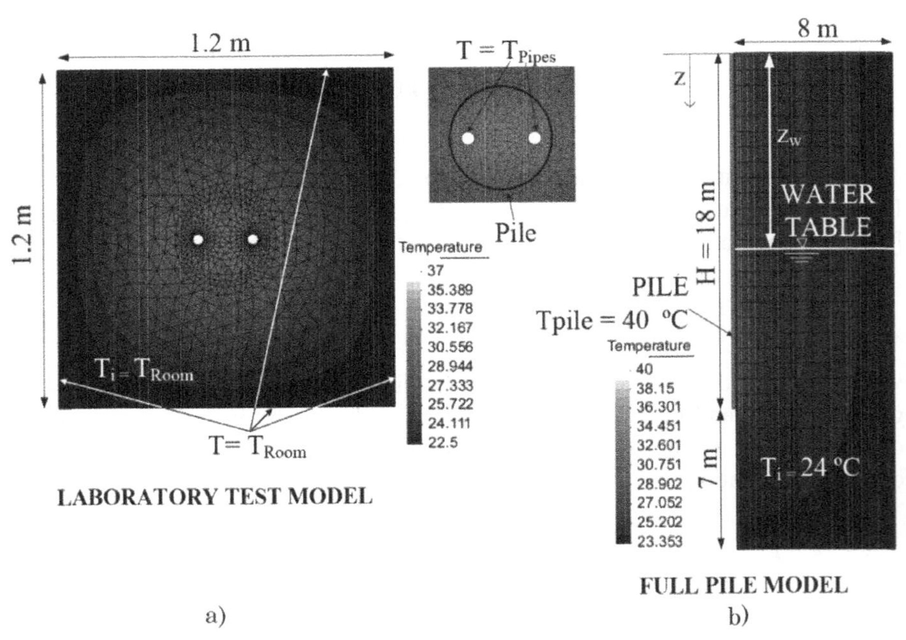

Figure 9.4. *Numerical models: a) the laboratory test model and b) the full pile model*

The planar section numerical model was solved for three soil conditions, as presented in Table 9.2 (Test 1, Test 4 and Test 6). The full pile numerical model was solved for three different soil types: sand, silt and sandy clay, with the soil water retention curve for the three soil types, as shown in Figure 9.5.

From the data shown in Figure 9.5, van Genuchten model parameters were estimated (Table 9.3). The van Genuchten model is shown in the same figure. In the full pile numerical model, for each soil type, five different positions of the water table (z_w/H = 0.00 (fully saturated), 0.25, 0.5, 0.75 and 1.0) were analyzed that resulted in five different soil saturation profiles.

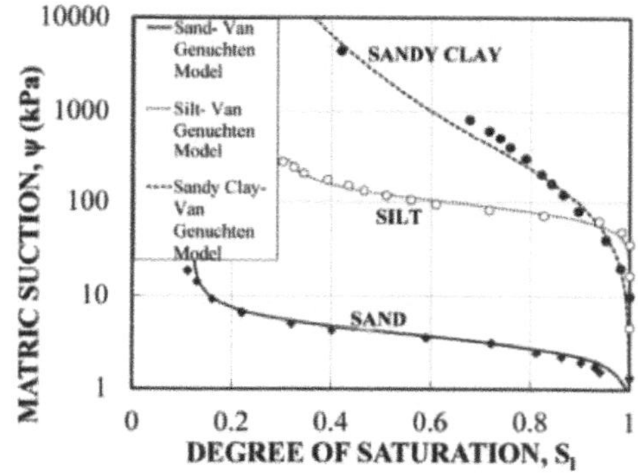

Figure 9.5. *Soil water retention curve for soils used in the full pile numerical model*

Soil type	P_0 (kPa)	λ_0	S_{rl}	ϕ
Sand [CLA 96]	3.4	0.75	0.122	0.38
Silt [BRO 64]	90	0.78	0.303	0.46
Sandy clay [VAN 99]	120	0.2	0.050	0.34

Table 9.3. *van Genuchten model parameters for sand, silt and sandy clay*

9.4.3. *Laboratory test and numerical results*

The main measurements from the laboratory tests were the temperature variation at the soil–pile interface and in the soil (points B, C, D, E, F and G, Figure 9.3). The heat flux cannot be measured, but it can be calculated from the numerical model. Figure 9.6 shows the measured and calculated temperature at the pile–soil interface and in the soil for dry soil conditions (Test 1). Figure 9.6 shows good consistency between the measured data and the numerical model, which supports the use of the numerical model to evaluate the heat flux.

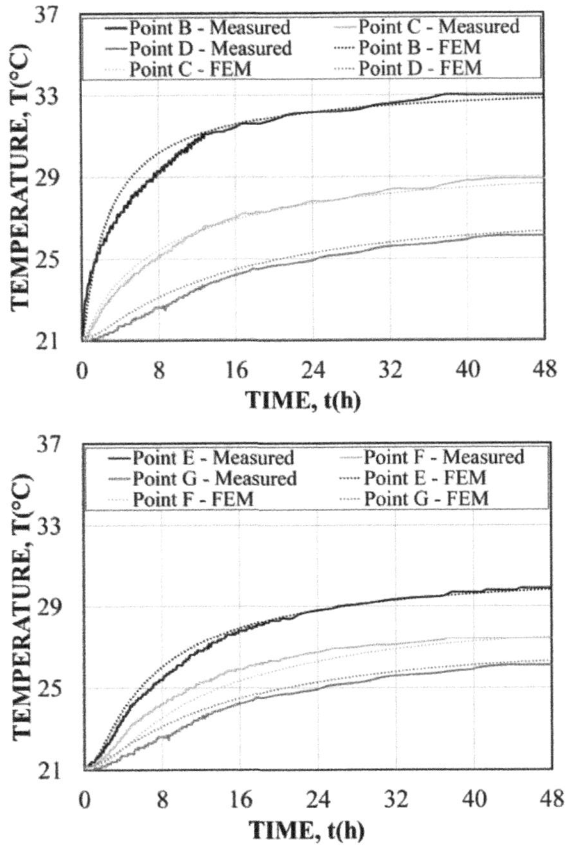

Figure 9.6. *Measured and calculated temperature at different location for Test 1*

The heat flux and temperature at the soil–pile interface for Test 1, 4 and 6 were evaluated from the numerical model and shown in Figure 9.7.

From Figure 9.7, it is clear that the heat flux was highly influenced by soil saturation conditions. In other words, the thermodynamic efficiency of the energy pile decreases in unsaturated soils. In addition, the heat flux and temperature was not uniform at the pile wall. It was maximum at the closest point to the pipes and minimum at the furthest point from the pipes. The total amount of heat exchanged with the soil per linear meter of pile, Q, is equal to the area below the heat flux curve of Figure 9.7. The measured normalized thermal conductivity and the efficiency of energy pile, ε, is shown versus S_l in Figure 9.8. Figure 9.8 shows that the efficiency of energy piles can decrease by 40% when the soil becomes dry.

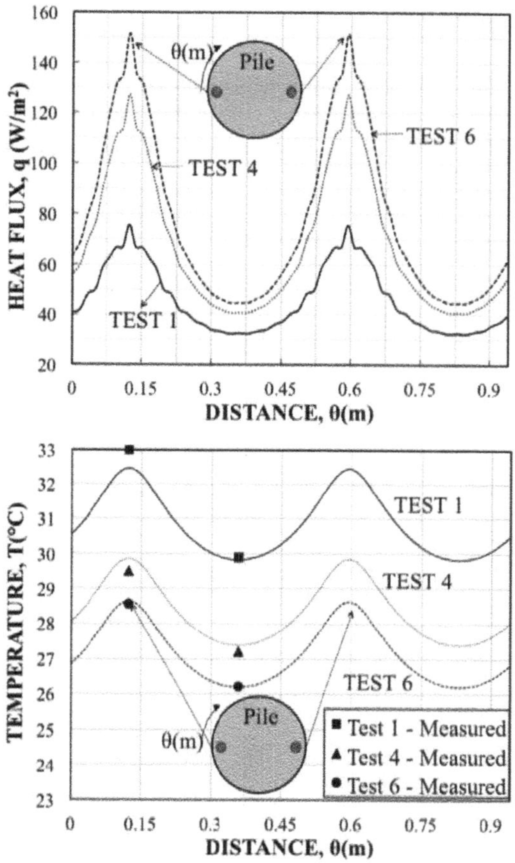

Figure 9.7. *Heat flux and temperature distribution at pile–soil interface*

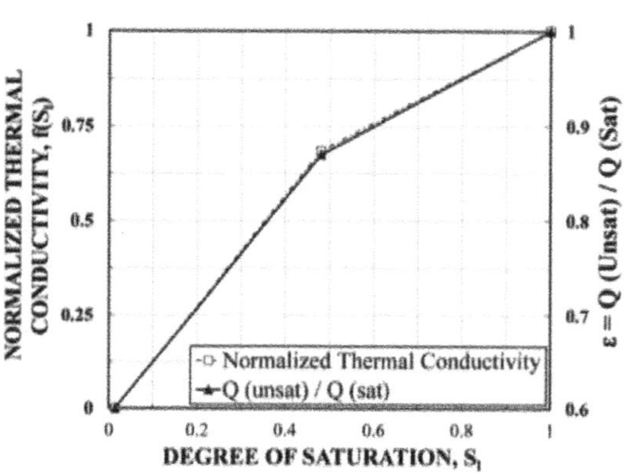

Figure 9.8. *Energy pile efficiency versus soil degree of saturation*

9.4.4. *Modeling the full pile*

The results of the full pile numerical simulation for three soil types and five water table positions (defined by z_w as in Figure 9.4) for each soil type are presented in Figure 9.9. The results include the soil saturation profile, S_l, and heat flux, q.

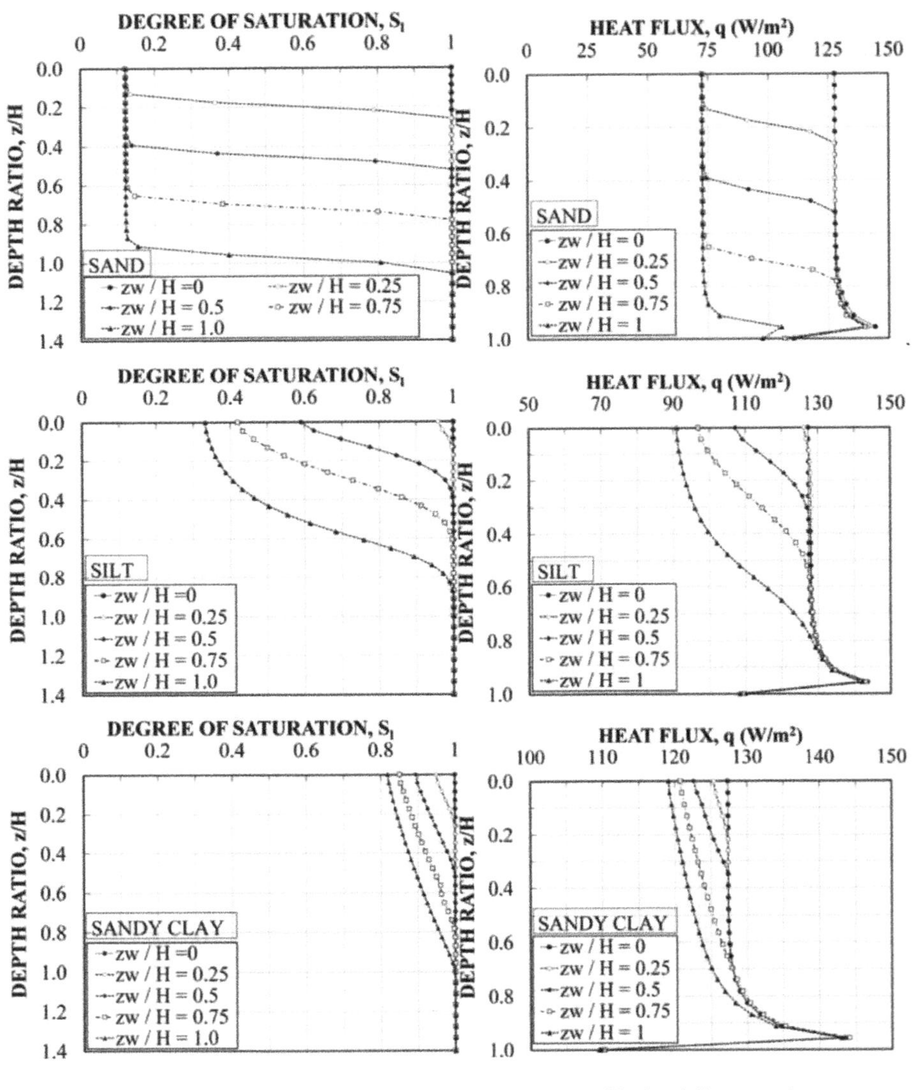

Figure 9.9. *Soil saturation and heat exchange profile for different soils*

The results presented in Figure 9.9 show that the heat flux was highly influenced by the soil saturation profile. The efficiency ratio for the overall soil profile heat exchange was evaluated from Figure 9.9 in the same manner as for the laboratory planar test. Figure 9.10 shows ε for the different soils with different soil saturation profiles.

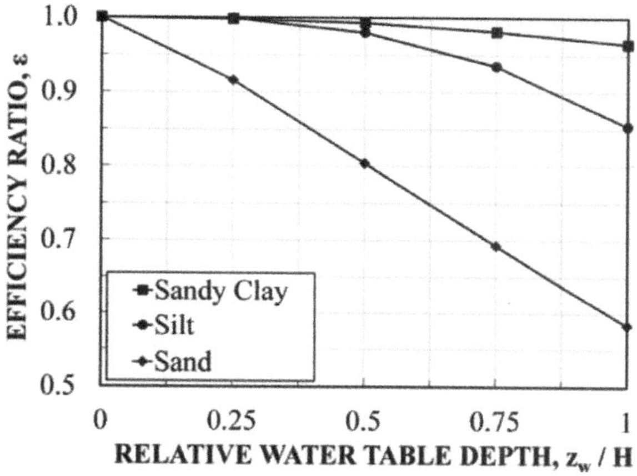

Figure 9.10. *Efficiency ratio for different soils under different conditions*

Figure 9.10 shows that the efficiency ratio, ε, varies with the soil saturation profile; and for the case studied in this chapter, where the water pressure is in hydrostatic conditions, the thermodynamic efficiency of an energy pile can decrease by 40% when the water table is at the bottom of the energy pile.

9.5. Full scale tests on energy piles

Three energy piles that are part of the Liberal Arts Building foundation were installed and instrumented as part of the ongoing research on energy geostructures at Texas A&M University. In addition, three separate boreholes were installed and instrumented with thermistors to measure and evaluate the temperature and heat transfer in the ground due to the operation of energy piles. The energy piles were connected to a geothermal heat pump, water tank and a water pump. The soil at the testing site consists of high-plasticity clay. Figure 9.11 shows the test setup.

The geothermal heat pump was operated for 24 h on a heating method resulting in cold water injection to the piles. After 24 h, the heat pump was turned off. The average inlet (T_{in}) and outlet (T_{out}) water temperature to the piles and the difference between them is shown in Figure 9.12. The resulting increase in pile P2 temperature at its center is also shown in Figure 9.12. After turning off the heat pump, the piles temperature starts increasing to reach the equilibrium with the soil temperature. Note that no change in soil temperature at any of the boreholes location was noticed during the operation time of the heat pump.

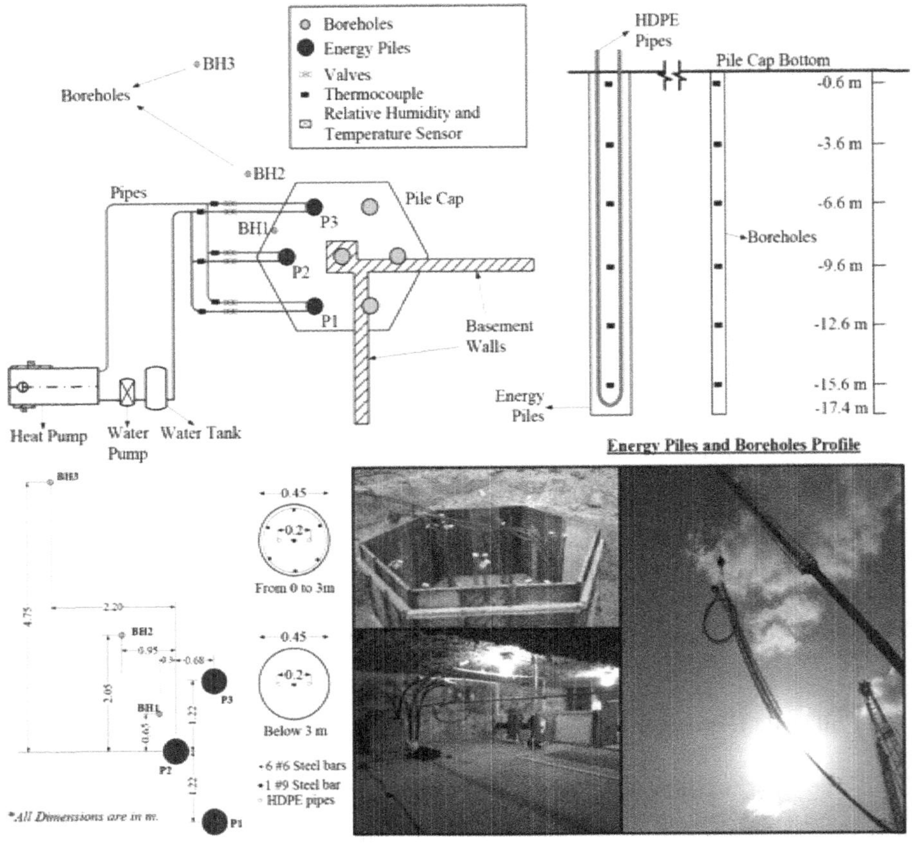

Figure 9.11. *Energy piles test setup and installation at Texas A&M University*

The initial and final piles' temperature after 24 h of heat pump operation, T1 and T2, respectively, is shown in Figure 9.13 where an average decrease of 10°C in pile temperature was measured.

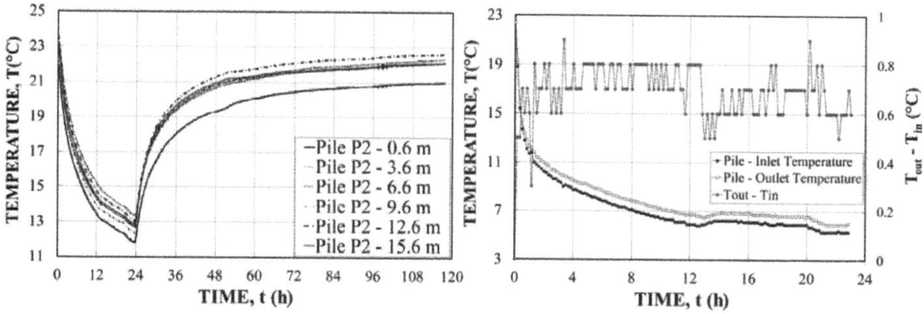

Figure 9.12. *Measured inlet water, outlet water, and pile P2 center temperature*

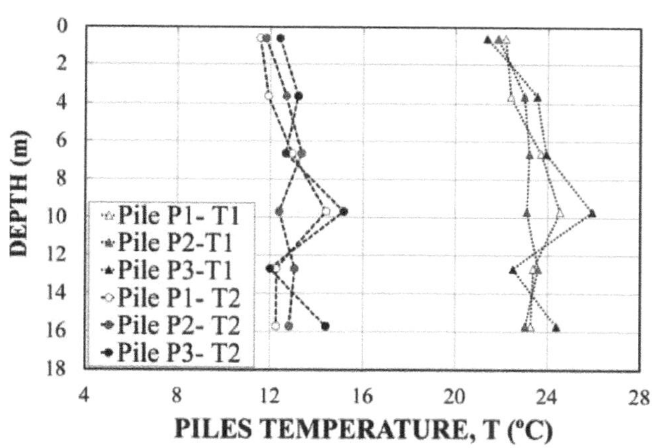

Figure 9.13. *Initial and final piles temperature*

9.6. Conclusions

This chapter presented a general overview on the impact of climatic factors on the performance of energy geostructures. This chapter also focused on the performance of energy piles in hot climates. Based on the results and information presented above, the sizing of energy geostructures should consider the climatic factors and the variation of soil conditions in the SSL throughout the year. In hot climates, the main impact of climatic factors on soil is variation of the soil saturation profile. Such variation directly impacts the mechanical and thermodynamic performance of energy geostructures, since a significant part of these structures lies in the variable SSL. The results and information provided in this chapter showed that the thermodynamic efficiency of energy piles can decrease by as much as 40% when considering the effect of climatic factors. In addition, the operation of the geothermal heat pump can significantly change energy geostructures' temperature resulting in additional stresses and strains.

9.7. Acknowledgments

The work presented in this chapter is supported by the Buchanan Chair at Texas A&M University, Texas A&M University system and Skanska General Contractors. The piles were installed and instrumented by Berkel & Company Contractors, the geothermal heat pump was donated and installed by TD Industries and the electrical equipment and work was donated and installed by Britt Rice Electric.

9.8. Bibliography

[ALO 99] ALONSO E., VAUNAT J., GENS A., "Modeling the mechanical behavior of expansive clays", *Engineering Geology*, vol. 54, pp.173–183, 1999.

[BRO 64] BROOKS R., COREY A., Hydraulic properties of porous media, Hydrology paper no. 3, Colorado State University, Fort Collins, CO, 1964.

[CLA 96] CLAYTON W.S., Relative permeability-saturation-capillary head relationships for air sparging in soils, PhD Dissertation, Colorado School of Mines, Golden, CO, 1996.

[CÔT 05] CÔTÉ J., KONRAD J.M., "A generalized thermal conductivity model for soils and construction materials", *Canadian Geotechnical Journal*, vol. 42, pp. 443–458, 2005.

[FOU 22] FOURIER J.B., *Théorie Analytique de la Chaleur*, Dover Publications, New York, NY, 1822.

[FRE 93] FREDLUND D.G., RAHARDJO H., *Soil Mechanics for Unsaturated Soils*, John Wiley & Sons, New York, 1993.

[GEN 98] GENS A., ALONSO E.E., "Constitutive models for unsaturated soils: thermodynamic approach", *Proceedings of the 2nd International Conference on Unsaturated Soils*, vol. 1, International Academic Publishers, Beijing, pp. 455–460, 1998.

[JOH 75] JOHANSEN O., Thermal conductivity of soils, PhD Thesis, Trondheim, Norway, 1975.

[LU 04] LU N., LIKOS W.J., *Unsaturated Soil Mechanics*, John Wiley & Sons, Inc., Hoboken, NJ, 2004.

[OLI 96] OLIVELLA S., GENS A., CARRERA J., *et al.*, "Numerical formulation for a simulator (CODE-BRIGHT) for the coupled analysis of saline media", *Engineering Computations*, vol. 13, no. 7, pp. 87–112, 1996.

[REE 00] REES S.W., ADJALI M.H., ZHOU Z., *et al.*, "Ground heat transfer effect of the thermal performance of the earth contact structures", *Renewable and Sustainable Energy*, vol. 4, no.3, pp. 213–265, 2000.

[SEP 79] SEPASKHAH A.R., BOERSMA L., "Thermal conductivity of soils as a function of temperature and water content", *Soil Science Society of America Journal*, vol. 43, pp. 439–444, 1979.

[SHA 47] SHANNON W.L., WELLS W.A., "Tests for thermal diffusivity of granular materials", *Proceedings of the American Society for Testing and Materials*, vol. 47, pp. 1044–1055, 1947.

[TAR 02] TARNAWSKI V.R., GORI F., "Enhancement of the cubic cell soil thermal conductivity model", *International Journal of Energy Research*, vol. 26, pp. 143–157, 2002.

[VAN 99] VANAPALLI S.K., PUFAHL D.E., FREDLUND D.G., "The influence of soil structure and stress history on the soil-water characteristic of a compacted till", *Géotechnique*, vol. 49, no. 2, pp. 143–159, 1999.

Chapter 10

Impact of Transient Heat Diffusion of a Thermoactive Pile on the Surrounding Soil

10.1. Introduction

Like every ground source heat pump (GSHP) system, thermoactive piles exchange heat with the ground using the earth as a heat source in the winter or a heat sink in the summer. The ground soil, wherein lies the geothermal energy, has a relatively steady temperature of about 10–15°C beneath the upper 6 m in most European countries [ADA 09]. Due to their particular seasonal energy demand to achieve an energy balance, thermoactive piles are subjected to seasonal thermal cycles throughout the year. The fluid circulating inside the absorber pipes contains a mixture of water and antifreeze (propylene glycol) [BRA 06], with a supply (inlet) fluid temperature that is usually set at 11–19°C above the steady ground temperature in the heating mode and at 6–8°C below the steady ground temperature in the cooling mode [KAV 10]. The thermal diffusion between the thermoactive piles and the ground soil generates a variation in the ground temperature profile. Because the ground temperature near the surface is greatly influenced by the air temperature and solar radiation, the ground temperature profile is not uniform in the uppermost few meters. Thus, the field of temperature diffusion will vary not only in transient seasonal time but also in depth. Furthermore, as the ground soil is a porous medium, the thermal variation consequently induces changes in its hydraulic and mechanical behaviors [FRO 99, LAL 99, BRA 06].

In this chapter, a brief study of the literature on heat transfer between the thermoactive piles and the surrounding soil is presented to achieve a better

---

Chapter written by Maria E. SURYATRIYASTUTI, Hussein MROUEH and Sébastien BURLON.

understanding of the complex coupled transfer. Several research studies on the thermal diffusion around the thermoactive piles have been reported with a combined heat conduction–advection analysis, taking into account the non-homogeneous soil conditions and the heat pump exchange rate [ROU 12, CHO 11]. This chapter presents a study that conducts a simplified numerical modeling of the transient heat conduction during the seasonal operation of a thermoactive pile without any coupling to the groundwater flow and the mechanical load of the building. The main objectives are to determine the temperature produced in the pile and at the soil–pile interface and to observe the change in ground temperature profile due to the seasonal temperature of fluid injected into the absorber pipes. Different types of absorber pipe and configuration of inlet–outlet flow were observed. Analyses are performed in a two-dimensional (2D) model at a given depth under a constant ground temperature as well as in a three-dimensional (3D) model under a sinusoidal ground temperature over the entire depth. The results give perspectives on the temperature-induced change in the mechanical and hydromechanical behaviors of the foundation, which depend strongly on the types of surrounding soil.

10.2. Heat transfer phenomenon

Heat transfer around the thermoactive piles and the surrounding soil strongly depends on the properties and mineral compositions of the porous material. If x is a porous material, x consists of a solid mineral s, a liquid f, and air a. Heat conduction occurs within the concrete piles and the soil grains, heat advection comes through the groundwater and moisture flow, vapor diffusion and latent heat transfer during the evaporation–condensation cycles of the groundwater might only be possible in partially saturated or unsaturated soil, and radiation at the soil surface contributes very little to the heat transfer [SUR 12].

The heat flux is transmitted principally by thermal conductivity λ and volumetric heat capacity C_v of the transfer medium. Volumetric heat capacity is the product of mass volume ρ with specific heat extraction c, which indicates the ability of a substance to store the heat energy. The ratio of thermal conductivity to the volumetric heat capacity is called thermal diffusivity, denoted as $D_h = \lambda/C_v$, referring to a value of the diffusion rate of a material. In the case of heat advection and vapor flux, hydraulic conductivity k, potential head ψ and latent heat L_v are the other major properties that govern the transfer. All these properties are strongly related to the mineral composition of the soil, the type and size of grains, the water and air content, as well as the porosity factor. Moist or saturated soil has a thermal capacity greater than that of dry soil because the thermal conductivity of dry soil is about five times lower than that of saturated soil [RIE 07]. Due to their larger grain size, granular soil has a hydraulic conductivity approximately three orders of magnitude greater than that of fine soil [DER 01]. The relations of heat conduction q_{cond}, heat advection q_{adv} and latent heat flux q_{latent} with the governed factors are expressed in equations [10.1]–[10.3]

$$\vec{q}_{cond} = -\lambda gradT \quad [10.1]$$

$$\vec{q}_{adv} = \rho_f c_f \vec{v}_{liq}\left(T - T_0\right) \quad [10.2]$$

$$\vec{q}_{latent} = L_v \vec{v}_{vap} \quad [10.3]$$

where T and T_0 denote the applied temperature and the initial temperature, respectively, and v_{liq} and v_{vap} indicate the groundwater velocity in the liquid phase and vapor phase, respectively.

10.2.1. *Soil properties*

10.2.1.1. *Ground temperature*

In the uppermost few meters, the ground temperature varies with time and depth at the heterothermal zone of the soil [BUR 85]. The variation in time *T(t)* is generally caused by external temperature (air temperature and solar radiation) while the propagation in depth *T(z)* is related to the heat flux and internal terrestrial heat in the soil. The variation in ground temperature forms a sinusoidal function, expressed in equation [10.4] [HIL 04].

$$T(z,t) = T_{ave} + A_0 e^{-\frac{z}{d}}\left(\sin\left(\omega t - \frac{z}{d}\right)\right) \quad [10.4]$$

where T_{ave} and A_0 denote the average annual ground temperature and the maximum annual amplitude of ground temperature, respectively, and d indicates the damping depth of the annual fluctuation, given by $d = (2D_h/\omega)^{1/2}$, where D_h is the thermal diffusivity of the soil and ω is the annual radial frequency ($2\pi/365$).

Fluctuation in ground temperature diminishes with depth and becomes steady to the average ground temperature at the neutral zone (i.e. at a depth of 10–50 m below the surface) [BUR 85]. More than 50 m below the surface, the soil is classified into the homothermal zone. In this zone, the ground temperature increases regularly with depth with a geothermal gradient of 3°C per 100 m [SAN 01].

10.2.1.2. *Groundwater flow*

The groundwater flow changes due to the variation of water content and temperature in both liquid and vapor phases. In general, the flow due to the water content gradient is greater than the flow arising from the temperature gradient, for both very dry (unsaturated) and very wet (saturated) media [PHI 57]. Moreover, the liquid diffusivity tends to play the most important role in saturated soil, while vapor diffusivity is dominant in unsaturated soil [PHI 57]. Groundwater flow can, therefore, be divided into the flow in saturated soil and the flow in unsaturated soil, expressed in equations [10.5] and [10.6], respectively.

$$\vec{v}_{liq} = -k\frac{\partial \psi}{\partial \theta} grad \ \ \theta = -k \ \ grad \ \ \psi \qquad [10.5]$$

$$\vec{v}_{vap} = -D_{\theta vap} \ grad \ \ \theta \qquad [10.6]$$

where θ and ψ indicate the water content and potential head, respectively, k represents the hydraulic conductivity and $D_{\theta vap}$ denotes the isothermal vapor diffusivity.

10.2.2. *Energy conservation in the transient regime*

To satisfy the energy conservation in the transient condition, the law of thermodynamics states that the time rate of change in the heat content of a volume element must be equal to the change of flux with the distance. If the surrounding soil has a substantial groundwater flow, the heat advection transfer cannot be ignored. The energy conservation should take into account the soil hydraulic gradient, given in equation [10.7] for saturated soil and in equation [10.8] for unsaturated soil.

$$C_v\frac{\partial T}{\partial t} - \rho_f c_f k \ \ grad \ \ \psi \ \ grad \ \ T = div \ \ (\lambda \ \ grad \ \ T) \qquad [10.7]$$

$$C_v\frac{\partial T}{\partial t} - \rho_f c_f k \ \ grad \ \ \psi \ \ grad \ \ T = div \ \ (\lambda \ \ grad \ \ T) - L_v \ \ div \ \left(D_{\theta vap} \ \ grad \ \ \theta\right) \qquad [10.8]$$

If there is no groundwater flow in the surrounding soil, the heat transfer in the thermoactive pile system is limited to pure heat conduction, with an addition of vapor transfer only in partially saturated or unsaturated soil [SUR 12]. Equations [10.9] and [10.10] are used to solve the energy conservation in saturated and unsaturated soil, respectively.

$$C_v\frac{\partial T}{\partial t} = div \ \ (\lambda \ \ grad \ \ T) \qquad [10.9]$$

$$C_v\frac{\partial T}{\partial t} = div \ \ (\lambda \ \ grad \ \ T) - L_v \ \ div \ \left(D_{\theta vap} \ \ grad \ \ \theta\right) \qquad [10.10]$$

10.3. Numerical modeling of thermal diffusion in a thermoactive pile

To study the diffusion of a thermoactive pile in its surrounding soil, numerical modeling using the finite element and finite difference methods is conducted. Analysis is performed in the transient regime during the seasonal operation time. Two study cases are proposed:

– a 2D model to observe the internal diffusion in the thermoactive pile itself (from the absorber pipes to the concrete pile);

– a 3D model to observe the diffusion around the thermoactive pile and the surrounding soil.

A concrete pile having diameter 60 cm and length 15 m is located in the center of a soil mass. The soil mass boundary dimension is selected after a parametric analysis to ensure an adiabatic boundary condition, with the horizontal radius set at 15 m and the height of the soil mass set at 30 m. The soil mass is assumed to be fully saturated sandy soil with no groundwater flow so that only heat conduction transfer is taken into account. The mesh is refined around the pile in order to increase the precision in the high strain gradient areas. The average annual ground temperature T_{ave} is set at 14°C. Four U-shaped absorber pipes are located 10 cm from the outer surface of the concrete pile, surrounding the pile with a 28-mm outer diameter. The position of the absorber pipes forms an anti-symmetric *I-O-I-O* configuration, where *I* designates the inlet flow and *O* the outlet flow. Figure 10.1 shows the cross-section detail of the thermoactive pile and Table 10.1 summarizes the thermal properties of the materials used in this study.

		Absorber pipes	Concrete pile	Soil
Density	ρ	960 kg/m^3	2,500 kg/m^3	1,950 kg/m^3
Thermal conductivity	λ	0.42 W/m°C	1.8 W/m°C	1.5 W/m°C
Specific heat extraction	c	2,000 J/kg°C	880 J/kg°C	800 J/kg°C
Coefficient of thermal expansion	α	1.4×10^{-4}/°C	1.2×10^{-5}/°C	5.0×10^{-6}/°C

Table 10.1. *Thermal properties of the materials used in the study [HIL 04]*

When the heat pump operates, the temperature of the heat carrier fluid in the inlet and outlet pipes generally varies for very short periods (i.e. a few hours) in accordance with the heat pump power. As a part of this ongoing research project, an experimental test of micro-thermoactive piles was conducted by ECOME from February 2011 to January 2012. The results showed that the delta temperature ΔT between the supply (inlet) fluid and the return (outlet) fluid is constant for one period of loading, explaining the good performance of the heat pump (Figure 10.2).

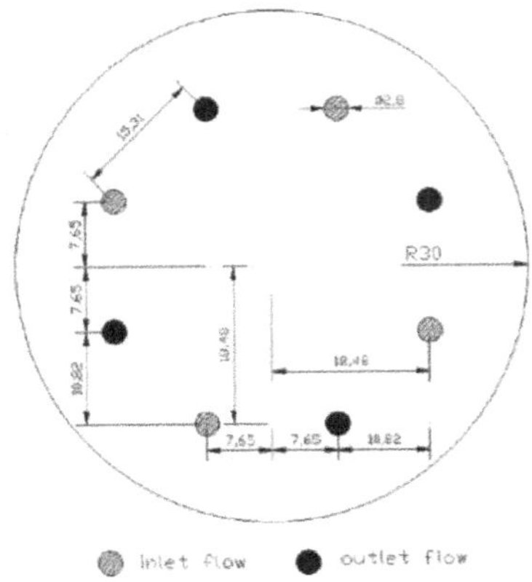

Figure 10.1. *Cross-section of the thermoactive pile*

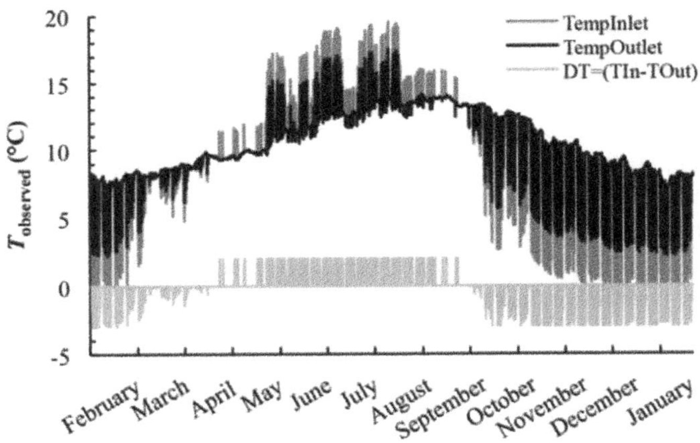

Figure 10.2. *Observation of the temperature in the micro-thermoactive piles by ECOME*

10.3.1. *A two-dimensional model – internal diffusion in the thermoactive pile*

The model takes the viewpoint of a plane surface at a certain depth z of a thermoactive pile system in a cylindrical coordinate. The absorber pipes are modeled as circular line sources inside a circular solid concrete pile in the center of the soil mass. Using this method, heat transfers can be evaluated in the radial direction, where differential equation of energy conservation is governed by the following differential equation:

$$C_v \frac{dT}{dt} = \lambda \left(\frac{d^2T}{dr^2} + \frac{1}{r}\frac{dT}{dr} \right) \qquad [10.11]$$

First, the analysis is performed in the winter mode with $T_{inlet} = 4°C$. Considering the energy performance of the heat pump, ΔT is set constant at −3°; thus, T_{outlet} is set at 7°C. Because of geometric limitations, the initial ground temperature is considered equal to the average annual ground temperature and constant with time and depth $T_{ini}(z,t) = T_{ave}$. Figure 10.3 shows the radial distribution of the temperature in the thermoactive pile at the end of 3 months of loading. After 3 months of loading, the temperature in the thermoactive pile is relatively uniform. The values of temperature produced in the pile center and at the pile perimeter are close to the average temperature of the absorber pipes.

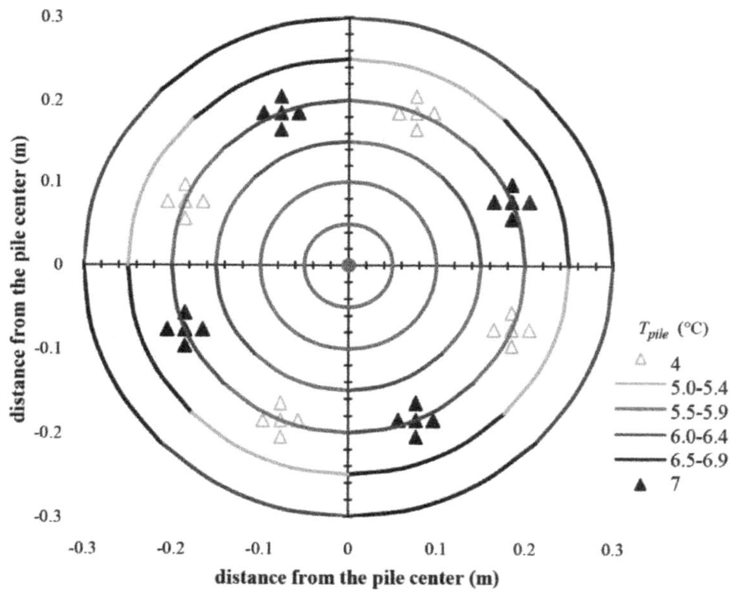

Figure 10.3. *Contour of temperature diffusion at the end of 3 months of loading*

10.3.1.1. *Effect of the absorber pipe configuration*

To study the influence of the absorber pipes in the heat diffusion, this study compared three types of absorber pipe configuration. Model A is the reference model as explained in the preceding section. Model B has four U-shaped absorber pipes, like model A, with a symmetric position in the *I-O-O-I* configuration. As for model C, two W-shaped absorber pipes were used with an antisymmetric *I-O-I-O* configuration.

With the same flow rate applied in each model, the results showed that the antisymmetric configuration results in a more uniform pile temperature than the symmetric configuration (Figure 10.4(a)). Model C with two W-shaped absorber pipes produced the highest pile temperature (Figure 10.4(b)), concluding that the W-shaped type is more effective than the U-shaped type in terms of energy productivity.

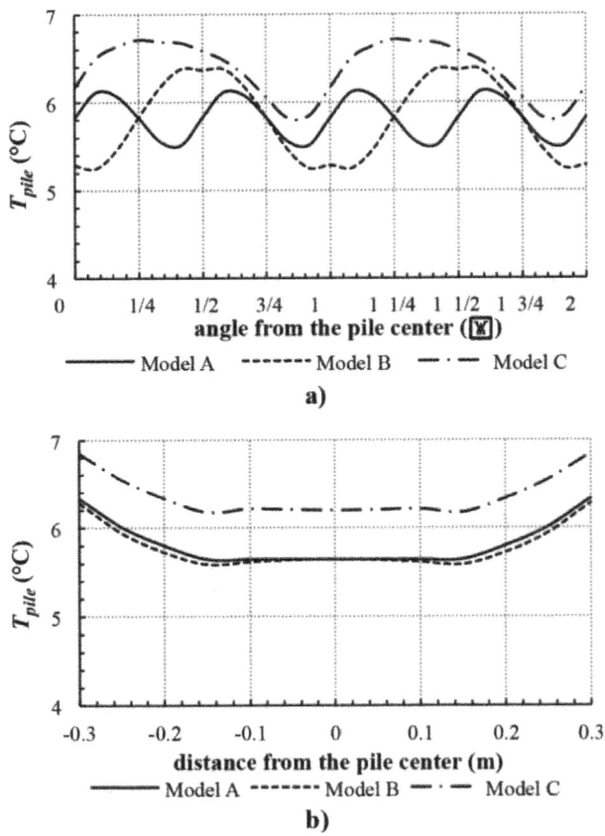

Figure 10.4. *Comparison of temperature profile in the pile: a) T(θ) and b) T(r)*

10.3.1.2. *Effect of different supply temperature*

Diffusion in seasonal operation time was analyzed with discontinuous thermal loads for every 3 months of loading. During each seasonal operation time, ΔT between the inlet flow and outlet flow was fixed, with respect to the experimental results, in order to maintain the energy efficiency of the heat pump. The design of seasonal fluid is illustrated in Figure 10.5 with ΔT set at −3°C in the cooling phase and +2°C in the heating phase. For each seasonal loading, the pile temperature varies progressively from the initial ground temperature and then steadily equilibrates to the average temperature of the absorber pipes (Figure 10.6).

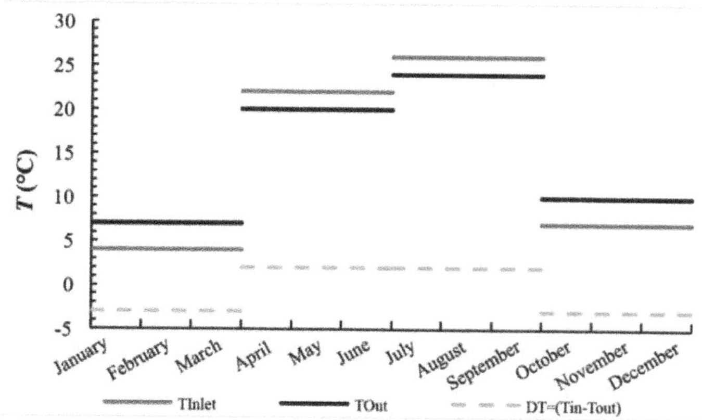

Figure 10.5. *Design of thermal loading over 1 year of operation time*

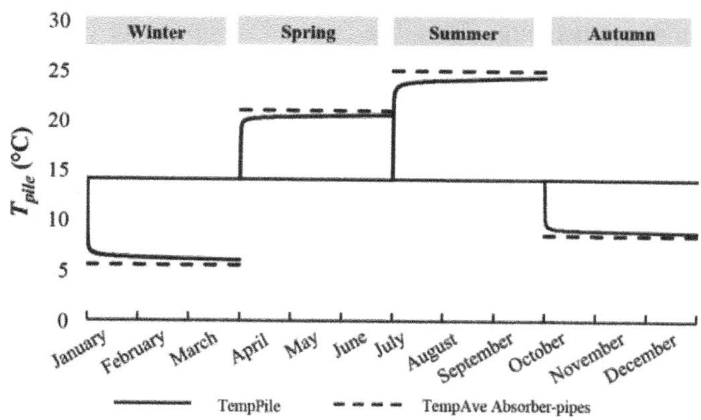

Figure 10.6. *Heat diffusion for each seasonal loading*

10.3.2. *A three-dimensional model – external diffusion to the surrounding soil*

The 3D numerical model is advantageous for modeling complex geometry with various combined loads. Accordingly, the variation of ground temperature in depth and time $T(z,t)$ (equation [10.4]) is taken into account in this study. Analysis is carried out in a coupled thermomechanical model using the finite difference code FLAC^{3D}. Because of the symmetry of the problem (load and geometric symmetry), only one-fourth of the complete domain is modeled. Consequently, a null flux is imposed on the symmetrical axes.

First, the ground soil is subjected initially to a sinusoidal surface temperature with an average ground temperature, T_{ave}, of 14°C. The soil mass has an internal terrestrial heat volume of 0.001 W/m^3 and a terrestrial surface flux of 0.0544 W/m^2 at depth 30 m. Figures 10.7(a) and 10.8(a) show the ground temperature profile in depth and season at the initial condition. It is shown that the sinusoidal surface temperature produced fluctuation in the ground temperature in the upper 6 m which becomes steady close to the average ground temperature at a depth $z > 10$ m.

Then, the thermoactive pile is activated in the soil system. The fluid temperature circulating in the pile is applied uniformly along the pile axes and is constant during one season of loading. The seasonal temperature values of the thermoactive pile are taken from the results of the 2D modeling, which are equal to the average temperature of the absorber pipes. The system is then subjected to combined sinusoidal surface temperature and seasonal thermoactive pile temperature. Figures 10.7(b) and 10.8(b) show the ground temperature profile after the activation of the thermoactive pile. It can be seen that the ground temperature equilibrium is quite disturbed due to seasonal thermal loading in the pile. Accordingly, the ground temperature at a depth $z > 10$ m for each season is no longer close to the average annual ground temperature T_{ave}. The ground temperature at a distance of approximately one pile diameter from the pile center has completely changed over the entire depth of a pile length.

10.4. Impact of the long-term thermal operation

10.4.1. *Groundwater flow effect on the heat diffusion*

With the presence of groundwater, the thermoactive pile system becomes a coupled thermohydromechanical system. The mass equilibrium of the soil between the groundwater velocity v (hydraulic) and the pore pressure p (mechanical) in relation to the soil–pile temperature gradient ΔT (thermal) is expressed in equation [10.12].

$$\beta_x \frac{\partial p}{\partial t} - \alpha_x \frac{\partial T}{\partial t} + div \ \vec{v} = 0 \qquad [10.12]$$

where x indicates the mineral composition of the soil related to its porosity factor and β and α denote the coefficient of compressibility and thermal expansion, respectively.

The presence of groundwater in the soil increases the thermal properties of the soil, including thermal conductivity and specific heat extraction up to 15% of those in the soil without groundwater [SAN 01]. Indeed, soil with a sufficiently large groundwater flow (up to 35 m/year) can produce natural heat energy regeneration [RIE 07, FRO 99]. In this case, the amount of heat extracted during winter no longer depends on the amount of heat injected in the summer. Consequently, the equilibrium of ground temperature is guaranteed; thus, the surrounding soil is not affected by the thermal volumetric variations. This case is of advantage to maintain the soil energy balance and minimize the operation cost of heat pump. In contrast, the concentration of heat energy is found around the piles and it will, therefore, modify the mechanical behavior of the concrete pile over long-term thermal operation [FRO 99].

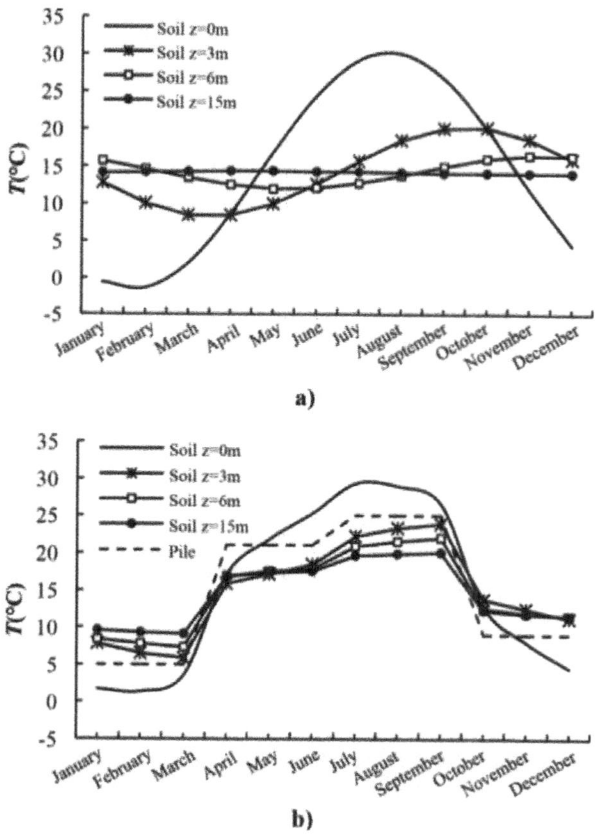

Figure 10.7. *Temperature profile T(t) at different depths: a) initial condition without the thermoactive pile and b) condition after the thermoactive pile is activated*

Due to the pore size, clayey soil generally has a low permeability; so, the groundwater velocity in clays is relatively low. Therefore, the heat advection transfer by groundwater flow is insignificant compared to the heat conduction in the soil grains. As a result, clays will undergo a significant temperature gradient that leads to an additional volumetric deformation and thus causes a variation in the pore pressure (see equation [10.12]). Referring to the mechanical equilibrium state in the soil, the change in pore pressure induces change in the effective stress simultaneously (see equation [10.13]).

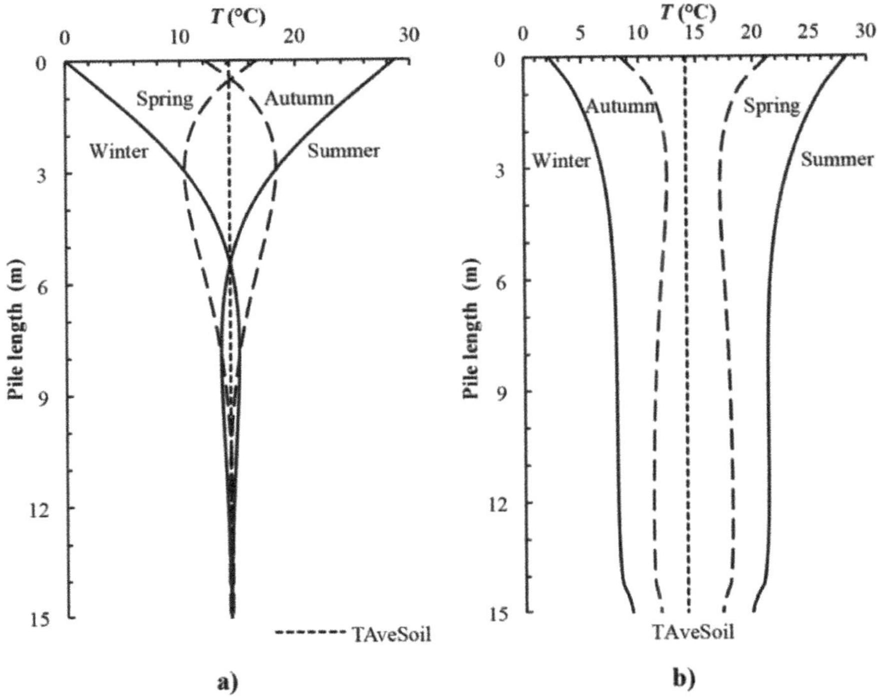

Figure 10.8. *Temperature profile T(z) at 0.6 m from the pile center: a) initial condition without the thermoactive pile and b) condition after the thermoactive pile is activated*

$$div\ \sigma - grad\ \ p + \rho g = 0 \qquad [10.13]$$

Regarding the thermoactive pile system, this condition is favorable; both the pile and the soil are subjected to thermal deformations; so, the relative displacements between the soil and the pile are small. Hence, the temperature-induced additional stress at the soil–pile interface is not remarkable. However, the temperature variation in the soil mass may change the soil properties and produce cyclic settlement and cyclic swelling of the mass and, thus, degrade the soil resistance [CEK 04].

Because sandy soil has a high permeability, the groundwater will rapidly diffuse the imposed pile temperature through a heat advection transfer phenomenon. In this case, the thermal gradient in sandy soil is very small. The soil will not undergo additional thermal volumetric deformation; only the concrete piles will undergo such a deformation. Hence, this is the unfavorable case scenario for the thermoactive pile system; temperature variations in the pile generate additional thermal stress in the concrete pile (see equation [10.14]), where a compressive force (resulting from the normal stress σ_{pile}) is taken as positive according to the convention in soil mechanics. As a result, the mobilized friction at the soil–pile interface will change in accordance with the additional thermal displacement [BOU 09, LAL 06] (see equation [10.15]).

$$\sigma_{pile} = E\left(\varepsilon + \varepsilon^{th}\right) = E\left(\varepsilon - \alpha\Delta T\right) \qquad [10.14]$$

$$EA\frac{d^2w}{dz^2} - \pi D \ q_s = 0 \qquad [10.15]$$

10.4.2. *Mechanical durability under thermal cyclic stress*

The numerical model discussed in section 10.3.2 represented the unfavorable case scenario in the thermoactive pile system. The modeled thermoactive pile underwent all of the additional stress induced by the temperature variations. The study considered a pure thermal loading with no service load applied from the upper structure, in an attempt to fully understand the effect of temperature on the soil–pile behavior.

Figure 10.9(a) shows the thermal axial displacement at different depths in the pile, whereas Figure 10.9(b) shows the soil surface settlement from the center of the pile. For the sake of clarity, the term "settlement" is designated by a positive value and "heave" by a negative value, with respect to the convention in soil mechanics. Figure 10.10 shows the non-uniform thermal deformation of the pile due to the ground temperature variations in depth and time. The results of temperature-induced mechanical response of a pile over 1 year of thermal operation confirmed the cyclic variation in the pile behavior. This cyclic response needs to be considered in the geotechnical capacity design of the thermoactive piles foundation. Moreover, during long-term thermal operation cycles, cyclic degradation in load–settlement capacity in the pile and mobilized shaft resistance at the soil–pile interface could develop.

10.5. Conclusions

Numerical analyses of transient heat diffusion of a thermoactive pile have been performed in this study. After a brief summary of the coupled transfer in the thermoactive pile system and the ground soil, the study is limited to heat conduction transfer analysis, considering a thermoactive pile in a surrounding sandy soil. The 2D analysis was capable of determining the temperature produced in the pile and at the soil–pile interface due to the fluid injected into the absorber pipes. The results showed that the pile temperature over the entire section is homogeneously close to the average temperature of the absorber pipes. The influence of various types and configurations of the absorber pipe was also analyzed. The W-shaped type produced a higher heat energy than the U-shaped type and the antisymmetric configuration produced the most uniform and homogeneous thermal diffusion inside the pile.

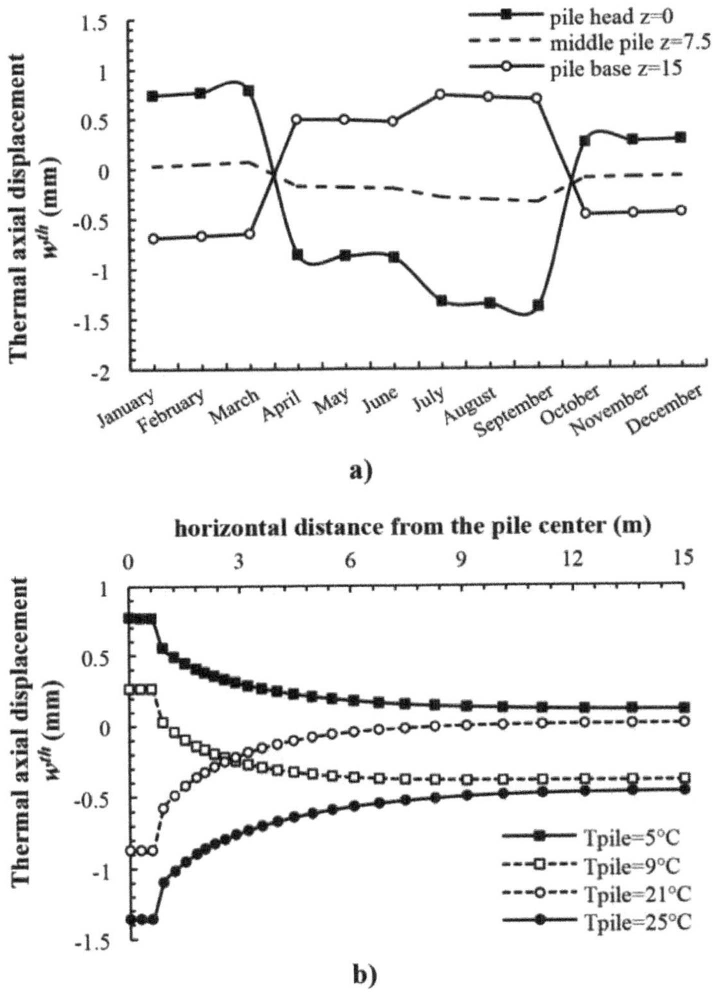

Figure 10.9. *Thermal axial displacement: a) in the pile at different depths and b) at the soil surface in different seasons*

The variation in the ground temperature with time and depth is taken into account in the 3D analysis. After installation of the thermoactive pile, the ground temperature profile in time and depth is varied considerably. Consequently, the average steady ground temperature at the soil neutral zone moved to a deeper zone. Depending on the types of soil (i.e. granular or fine soil), groundwater plays an important role in the heat diffusion and, therefore, controls not only the ground temperature equilibrium, but also the soil–pile stress equilibrium. Hence, a coupled thermo-hydro-mechanical analysis is required to achieve a better understanding of the impact of long-term thermal operation in the thermoactive piles. Also, consideration of cyclic response in the soil and concrete stress should be taken into account to enhance the design and dimensioning of this pile foundation.

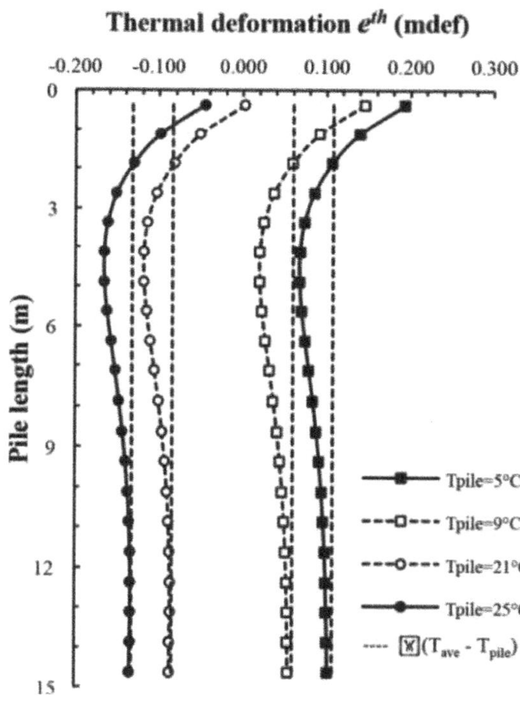

Figure 10.10. *Thermal deformation in the thermoactive pile*

10.6. Acknowledgments

The work described in this chapter is a part of the research project GECKO (*geostructures and hybrid solar panel coupling for optimized energy storage*), which is supported by a grant from the French National Research Agency (ANR). It is an industrial project with international collaboration, involving companies and research laboratories in civil and energy engineering sector: ECOME, BRGM, IFSTTAR,

CETE Nord Picardie, LGCgE-University of Lille 1, LEMTA-INPL and EPFL. The authors would like to express their gratitude to their partners for the valuable discussions and exchanges in this ongoing project.

10.7. Bibliography

[ADA 09] ADAM D., MARKIEWICZ R., "Energy from earth-coupled structures, foundations, tunnels and sewers", *Géotechnique*, vol. 59, no. 3, pp. 229–236, 2009.

[BOU 09] BOURNE-WEBB P.J., AMATYA B., SOGA K., *et al.*, "Energy pile test at Lambeth College, London: geotechnical and thermodynamic aspects of pile response to heat cycles", *Géotechnique*, vol. 59, no. 3, pp. 237–248, 2009.

[BRA 06] BRANDL H., "Energy foundations and other thermo-active ground structures", *Géotechnique*, vol. 56, no. 2, pp. 81–122, 2006.

[BUR 85] BURGER A., RECORDON E., BOVER D., *et al.*, *Thermique des nappes souterraines*, Presses Polytechnique Romandes, Lausanne, 1985.

[CEK 04] CEKEVARAC C., LALOUI L., "Experimental study of thermal effects on the mechanical behavior of a clay", *International Journal for Numerical and Analytical Methods in Geomechanics*, vol. 28, no. 3, pp. 209–228, 2004.

[CHO 11] CHOI J.C., LEE S.R., LEE D.S., "Numerical simulation of vertical ground heat exchangers: intermittent operation in unsaturated soil conditions", *Computers and Geotechnics*, vol. 38, pp. 949–958, 2011.

[FRO 99] FROMENTIN A., PAHUD D., LALOUI L., *et al.*, "Pieux échangeurs: Conception et règles de pré-dimensionnement", *Revue française de génie civil*, vol. 3, no. 6, pp. 387–421, 1999.

[HIL 04] HILLEL D., *Introduction to Environmental Soil Physics*, Elsevier Academic Press, Amsterdam, 2004.

[KAV 10] KAVANAUGH S., Ground source heat pump system designer – an instruction guide for using a design tool for vertical ground-coupled, groundwater and surface water heat pumps systems, Energy Information Services, report, Northport, AL, 2010.

[LAL 99] LALOUI L., MORENI M., FROMENTIN A., *et al.*, "Heat exchanger pile: effect of the thermal solicitations on its mechanical properties", *Bulletin d'Hydrogéologie*, vol. 17, pp. 331–340, 1999.

[LAL 06] LALOUI L., NUTH M., VULLIET L., "Experimental and numerical investigations of the behaviour of a heat exchanger pile", *International Journal for Numerical and Analytical Methods in Geomechanics*, vol. 30, no. 8, pp. 763–781, 2006.

[PHI 57] PHILIP J.R., DE VRIES D.A., "Moisture movement in porous materials under temperature gradients", *Transactions – American Geophysical Union*, vol. 38, pp. 222–232, 1957.

[RIE 07] RIEDERER P., EVARS G., GOURMEZ D., *et al.*, Conception de fondations géothermiques, Centre Scientifique et Technique du Batiment, final report ESE/ENR no 07.044RS, Sophia-Antipolis, 2007.

[ROU 12] ROUISSI K., KRARTI M., MCCARTNEY J.S., "Analysis of thermoactive foundations with U-tube heat exchangers", *Journal of Solar Energy Engineering*, vol. 134, no. 2, pp. 021008-1–021008-8, 2012.

[SAN 01] SANER B., "Shallow geothermal energy", *GHC Bulletin*, vol. 22, pp. 19–25, 2001.

[SUR 12] SURYATRIYASTUTI M.E., MROUEH H., BURLON S., "Understanding the temperature-induced mechanical behavior of energy pile foundations", *Renewable and Sustainable Energy Reviews*, vol. 16, pp. 3344–3354, 2012.

Chapter 11

Ground-Source Bridge Deck De-icing Systems Using Energy Foundations

11.1. Introduction

Bridges are an integral part of the highway infrastructure and can be critical assets for a nation's economy and security. Deterioration of the aging bridge infrastructure is a major concern that presents significant economic and engineering challenges. For example, there are about 600,000 bridges in the United States, of which about 60% are built using either conventional reinforced concrete or prestressed concrete [FED 08]. The same report classifies a quarter of the bridges in the United States, including their substructures, as structurally deficient or functionally obsolete.

The icing of bridge decks in the winter is a potential problem that creates dangerous driving conditions for motorists. The current methods for de-icing bridge decks are energy intensive, corrosive to the bridge itself, and/or dangerous to the environment. Salts and other chemicals commonly used to de-ice concrete bridge decks in the winter lead to accelerated corrosion of the bridge structure, threatening its structural integrity, and can also have long-term life cycle impacts due to increased maintenance and repair costs. One of the major factors that can lead to accelerated deterioration of bridge infrastructure is the chloride attack from de-icing salts [CAD 83, BAB 91].

The deterioration of bridge infrastructure due to chloride attack from de-icing salts has been extensively documented [VIR 83, POU 04, WAN 06, GRA 09].

Chapter written by C. Guney OLGUN and G. Allen BOWERS.

Corrosion of steel reinforcement in concrete bridge decks can be detrimental to the structural integrity of the reinforced concrete bridge deck. The oxidization of the steel reinforcement through chemical attack reduces the cross-sectional area and overstresses the reinforcement. Several collapses of overpasses and bridges have been attributed to the corrosion of steel reinforcement within the bridge decks and reinforced concrete beams [NAI 10]. It is estimated that the direct annual cost of bridge corrosion in the United States ranges from $6 to $10 billion [KOC 02]. The actual total cost can be as much as 10 times higher than this estimate if the indirect costs are taken into the cost calculation [YUN 03].

There are several alternatives to chemical de-icing such as heating the bridge deck using electrical, boiler-powered and ground-source systems. Electrical systems utilize the heat generated from the electrical resistance when an electric current is applied to a wire in the deck. For de-icing applications, mineral-insulated electric cables are embedded within the bridge deck and the system is turned on whenever needed [HAV 78]. A second utilization of electrical heating is electrically conductive concrete (ECC) which is produced by adding electrically conductive steel fibers and carbon particles to the concrete mix. When a current is applied, the electric resistance heats the entire slab of concrete [YEH 00, TUA 04, TUA 08]. Carbon fiber heating wire (CFHW) also uses electrical resistance to produce heat energy. In one study, the effectiveness of snow melting on a concrete slab using CFHW has been successfully demonstrated [ZHA 10]. However, this technology has not yet been implemented in full scale. Boiler-powered systems heat the fluid and circulate it mechanically (hydronic system) or convectively (heat pipe) through circulation tubes embedded within the bridge deck. Several experimental studies have been reported where boiler-heated hydronic systems are utilized [MIN 99]. A full-scale case study of a bridge that was fitted with a heat pipe system is also reported [HOP 01].

Ground-source heating can be used to de-ice bridge decks and results in a significant reduction in the use of salts and chemicals. In this concept, the constant temperature of the ground and its thermal storage capacity can be utilized as a renewable heat source that can be exploited for heating bridge decks in the winter. This technology can potentially eliminate or drastically reduce the need for chemical de-icing agents. It can also be utilized to reduce the temperature of the bridge deck during concrete curing and help minimize early age cracking. Similarly, it can be used to regulate the temperature of the bridge deck and reduce the severity of heating/cooling cycles between the day and the night in the summer. To accurately assess the feasibility of using energy foundations to control the temperature of the bridge deck, several considerations must be addressed. They include: the energy transfer mechanisms and mechanical properties of the bridge deck, the thermomechanical soil–pile interactions, and the coupled behavior of the system.

This chapter summarizes the thermal processes involved in the heating of bridge decks and presents the results of numerical analyses to provide more insights into the operational principles of bridge deck de-icing. A series of numerical analyses were performed to investigate the effect of system-related parameters, such as tube spacing, concrete cover thickness and fluid flow rate, as well as the effect of environmental factors, such as inlet fluid temperature, ambient temperature and wind speed. The results are presented and discussed to serve as a benchmark from which to measure the operational conditions and the energy requirements for designing ground-source bridge deck de-icing systems.

11.2. Ground-source heating of bridge decks

The ground can be utilized as a heat source that can be exploited for the ground-source heating of bridge decks in the winter. The heat energy in the ground can be harvested using energy foundations, geothermal boreholes or shallow trenches integrated with circulation tubes. Heat energy stored in the ground can be extracted and, in turn, circulated as warm fluid within the tubing system in the bridge deck to prevent icing [KUM 88]. In one particular application, warm water extracted from groundwater wells is sprinkled on the road surface for snow melting and de-icing [KAY 80].

A schematic of the concept is shown in Figure 11.1, where the energy piles and the approach embankment provide pathways for accessing the stored energy in the ground. Heat energy can be extracted from the ground as fluid is circulated within the tubing system in heat exchanger elements. Fluid circulation can be performed using either a heat pump or a geothermal circulation pump. Heat pumps are more efficient in heat exchange with the ground and can deliver higher temperature levels to the bridge deck [LIU 07]. However, the operation of heat pumps can be energy intensive and requires the availability of an external power source. Therefore, these types of systems are generally limited to large-scale applications and/or cases with significant heating energy demands. On the contrary, geothermal circulation pumps require less electrical power, which can be supplied by a solar-power integrated battery system. In this case, the fluid extracted from the ground will have a temperature close to that of the ground and the system will operate at a much lower temperature than a heat pump system. Therefore, heating the bridge deck using a circulation pump can be considered a passive heating system where the average ground temperature serves as a baseline for the fluid temperature. This type of application imposes operational limitations on the bridge deck de-icing operations, where heat injection rates into the bridge deck slab will not be as high as heat pump systems.

The temperature in the bridge deck slab can rise in excess of 50°C during the daytime in the summer, followed by a significant drop due to heat loss and cooling during the night [IMB 85]. Cyclic strains induced by these thermal expansion/contraction episodes, or the stresses that can develop due to the physical restraining of the slab, can fatigue the bridge deck slab, resulting in reduced life span and increased maintenance. The tubing system can also be utilized to lower the temperature by running the ground-source system during the daytime in the summer. In fact, the bridge deck functions as an energy collection system and the harvested thermal energy can be injected into the ground in the summer. This operation not only reduces the excess temperature and cyclic straining of the bridge deck but also helps increase the ground temperature. As a result of the increased temperature, the ground can serve as a more efficient energy source for de-icing the bridge deck in the winter.

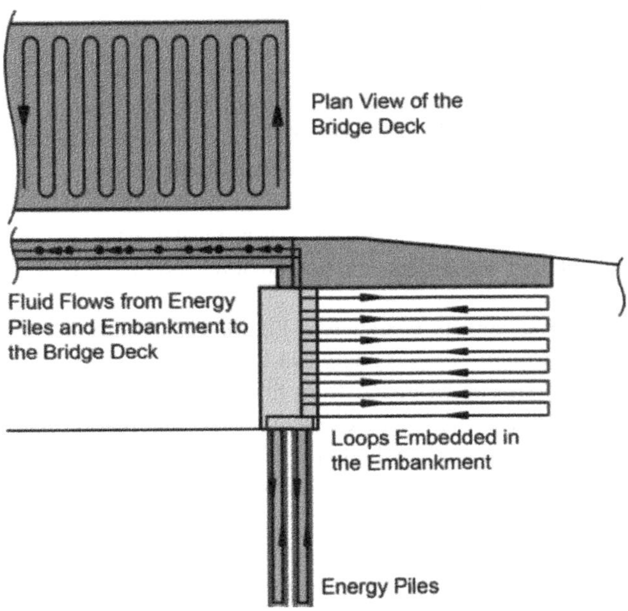

Figure 11.1. *Conceptual schematic of ground-source bridge deck de-icing*

11.3. Thermal processes and evaluation of energy demand for ground-source de-icing systems

Thermal processes involved in bridge deck de-icing are schematically shown in Figure 11.2. Heating of the bridge deck consists of several heat transfer mechanisms within the bridge deck as well as the thermal interaction of the bridge deck with the environment. Heat transfer modes during bridge deck de-icing include conduction, convection, radiation, and latent heat exchange during phase change of the precipitation [BER 11].

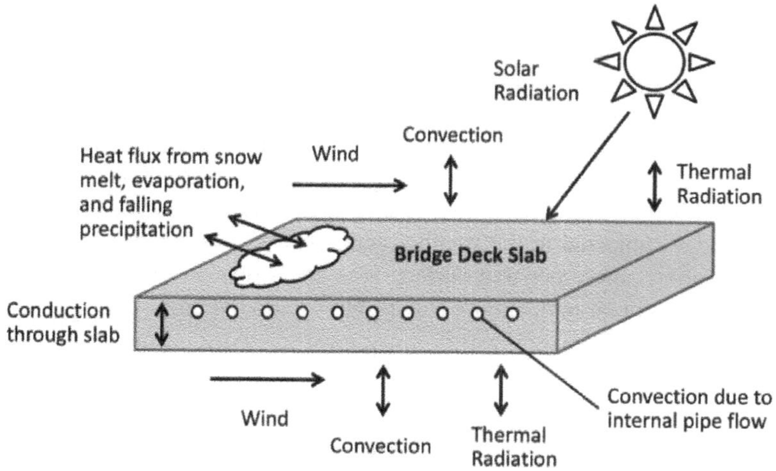

Figure 11.2. *Thermal processes in ground-source bridge deck de-icing, adapted from [SPI 00]*

Conduction is the predominant mode of heat transfer within the bridge deck. Conduction occurs within the circulation tube, throughout the concrete bridge deck and between the precipitation and the bridge deck. Convection is a flow-induced heat transfer mechanism within the bridge deck as the circulation fluid carries heat energy and transfers it to the bridge deck through the circulation tubes. Convection also takes place between the air and the bridge deck as wind moves the air and facilitates convective heat transfer between the deck and the outside environment. Radiation is also one of the significant energy transfer mechanisms between the bridge deck and the environment. The bridge deck absorbs solar radiation and long-wave radiation from the atmosphere and also emits long-wave radiation back to the atmosphere. The amount of radiation a bridge deck absorbs or emits is influenced by the temperature, cloud cover, the type of bridge deck surface, time of day and time of year. In addition to these processes, the melting and evaporation of snow precipitation also involves significant heat energy that affects the bridge heating process.

The energy demand of a bridge deck de-icing system can be estimated by accounting for the thermal processes described above. Chapter 51 of the American Society of Heating, Refrigeration and Air-Conditioning Engineers (ASHRAE) Applications Handbook presents a method for calculating the required heat flux for snow-melting operations [AME 11]. The inputs for this method include atmospheric conditions (ambient air temperature, amount of cloud cover, rate of snow fall, wind velocity, relative humidity, etc.), geometry of the slab, emittance factor of the concrete surface, and the desired ratio of snow-free area to total area.

This method considers several heat flux components to estimate the total heat flux needed for de-icing the bridge deck. The sensible heat flux is the heating power required to raise the temperature of the falling snow to melting temperature, and the latent heat flux is the heat flux required to melt the snow. The convective, radiative and evaporative heat fluxes are the heat fluxes required to account for energy loss due to wind, radiation and evaporation, respectively. As an example, heating energy demand calculations were performed using the ASHRAE method for a typical two-lane bridge subjected to the weather conditions in Blacksburg, Virginia, USA. Levels of heating power per unit surface area are shown in Figure 11.3 for given wind speed, ambient air temperature and a 1 cm/h snowfall rate.

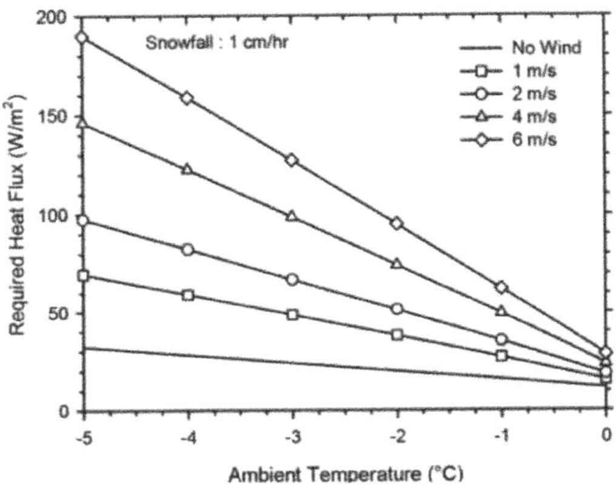

Figure 11.3. *Conceptual schematic of ground-source bridge deck de-icing*

11.4. Numerical modeling and analysis results

A series of three-dimensional numerical analyses were performed to model the bridge deck heating process using the finite element software COMSOL. The bridge deck slab modeled in the analyses is shown in Figure 11.4. The dimensions of the slab are 6.6 m × 4.5 m × 0.25 m. The hydronic piping selected was a 1.9 cm OD/1.3 cm ID PEX pipe. Warm fluid was circulated through the circulation tube at a constant flow rate and the temperature progression within the deck was evaluated. Water with 25% propylene glycol was used as the carrier fluid. The temperature of the inlet fluid is a reflection of the *in situ* ground temperature as these analyses considered the use of a circulation pump rather than a heat pump. Inlet fluid temperature was kept constant throughout the analyses even though this temperature is likely to vary slightly as a result of colder fluid being injected into the ground.

The analyses considered a variety of tube spacings, inlet fluid temperature, flow rate, wind speed, ambient temperature and thickness of concrete cover over the circulation tubes. The ambient temperature was kept constant throughout the analyses. A variation of these parameters allowed us to evaluate the respective effects of different factors on the bridge deck heating process. The rebar was not modeled in the analyses because relative to the concrete, its volumetric mass and heat capacity are small, and its thermal conductivity is much higher. In addition, the diameter of the rebar is very small relative to the dimensions of the slab and would require an extremely fine mesh, significantly increasing the computational time. Preliminary analyses have been performed with and without the rebar. It was concluded that the effect of the rebar was negligible.

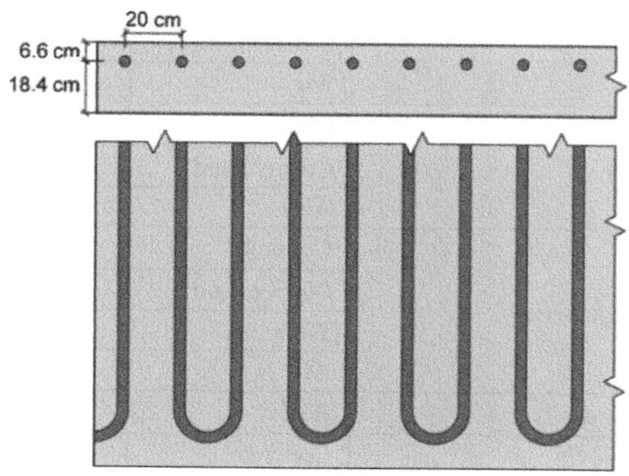

Figure 11.4. *Bridge deck slab used in the analyses and layout of the circulation tube*

The analyses were limited to the heating process of the bridge deck while the ambient temperature was kept constant and the melting of the snow was not included. The main purposes of these assumptions were to maintain simplicity in the calculation and also to develop a baseline for the energy demand required to bring the bridge deck to above freezing temperature. This condition can be considered as the case when the bridge deck surface is heated to an above freezing temperature preemptively before snowfall. In this case, the bridge deck will remain snow-free after precipitation if the heat injection compensates the latent heat from the melting of the snow after the start of precipitation.

Properties of the materials used in the analyses are summarized in Table 11.1. A total of 256 cases were analyzed, where different model parameters were systematically varied, as summarized in Table 11.2. The center–to–center spacing of the circulation tubes for the base case was set at 20 cm and the centerline of the tubes was 6.6 cm below the deck surface which corresponds to a concrete cover of 5.7 cm above the tubes as recommended [ACI 11]. The fluid with 12°C inlet temperature was circulated at a flow rate of 11.4 L/min. The initial slab temperature is −2°C, and no wind was considered for the base case. Temperature profiles along the vertical section at the mid-point between two tubes are shown in Figure 11.5 for different times into heating. This centerline section represents the most distant point from each tube in the horizontal direction. It can be seen that the temperature rise is quickest near the same elevation as the tube within the slab. The slab gets progressively warmer with higher temperature, as expected, at the surface in comparison to the slab base. In this analysis, the top 8.5 cm of the deck slab is reached above 0°C at the end of 1.5 h of heating using 12°C circulation fluid.

Property	Material	Value
Density	Concrete	2,408 kg/m^3
	Carrier fluid	1,041 kg/m^3
	Air	1.23 kg/m^3
Heat capacity	Concrete	880 J/kg·K
	Carrier fluid	3,691 J/kg·K
	Air	1,006 J/kg·K
Thermal conductivity	Concrete	1.44 W/m·K
	Carrier fluid	0.48 W/m·K
	Tube	0.41 W/m·K
	Air	0.0239 W/m·K
Kinematic viscosity	Air	1.315×10^{-5} m^2/s
Surface emissivity	Concrete	0.91
Dynamic viscosity	Carrier fluid	0.00273 kg/m·s
Prandtl number	Air	0.72

Table 11.1. *Summary of the material properties used in the numerical analyses*

Tube spacing (cm)	Wind speed (m/s)	Concrete cover (cm)	Inlet fluid temperature (°C)	Ambient temperature (°C)	Flow rate (L/min)	Number of runs
20	0	5.7	12	−2.0	11.4	Base case
15, 20, 25, 30	0	3.7, 5.7, 7.7, 9.7	6, 8, 10, 12, 14, 16, 18, 20	−2.0	11.4	127
15, 20, 25, 30	1, 2, 4	5.7	6, 8, 10, 12, 14, 16, 18, 20	−2.0	11.4	96
15, 20, 25	6	5.7	6, 8, 10, 12, 14, 16, 18, 20	−2.0	11.4	24
20	0	5.7	12	−0.5, −1.0, −1.5, −2.0, −2.5	11.4	4
20	0	5.7	12	−2.0	6, 8, 15, 20	4

Table 11.2. *Model parameters used in the numerical analyses*

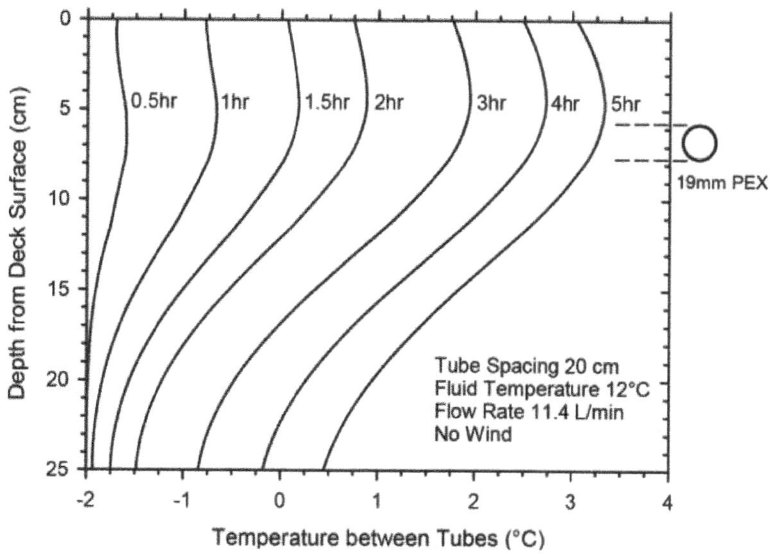

Figure 11.5. *Temperature along the vertical section in between circulation tubes*

Progression of temperature for the base case analysis is shown in Figure 11.6(a) as a result of fluid circulation with inlet temperature of 12°C at a flow rate of 11.4 L/min. This figure shows the deck surface temperature above the circulation tube and in between the tubes in comparison to the average surface temperature. It is seen that the surface temperature increases from the initial value of −2°C more rapidly above the tube location than the mid-point between the tubes, taking 0.71 and 1.47 h to reach 0°C at these locations, respectively. The effect of wind speed is demonstrated in Figure 11.6(b) where the progression of average deck surface temperature is shown for cases with various wind speed. As expected, the convective cooling effect of the wind inhibits the heating process as it removes heat from the bridge deck slab, while the ambient temperature remains constant. It takes 1.12, 1.26, 1.39, 1.68 and 2.06 h for an average deck surface temperature to reach 0°C, for conditions with 0, 1, 2, 4 and 6 m/s wind speed, respectively. On the contrary, it takes 1.81, 2.22, 2.67, 4.08 and 8.18 h to reach 1°C average deck surface temperature for the same wind conditions. It is also instructive to note that heating the bridge deck slab takes increasingly longer at higher wind speeds. This shows the significance of wind conditions during a de-icing process, which can be critical for determining the required ground-source heating capacity for a given performance level.

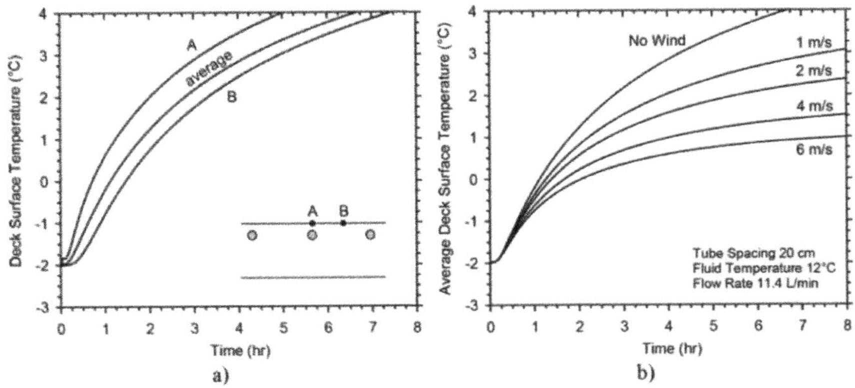

Figure 11.6. *Temperature increase at bridge deck surface in response to warm fluid circulation*

The rate of heat energy injected into the bridge deck slab can be calculated from the difference between the inlet fluid temperature and the temperature of the fluid exiting from the other end of the slab for a given flow rate. Injected heating power per unit surface area of the bridge deck is shown for various analyses where the flow rate was 11.4 L/min. Analyses are performed for a full range of inlet fluid temperatures, and the results are presented for certain cases. It is observed that the injected heat rate sharply increases during the first hour of the analyses and follows fairly stable values for each case. The heat injection rates for unit slab surface area, after the sharp rise, are 20, 28, 35 and 43 W/m^2 for 8, 12, 16 and 20°C inlet fluid temperatures, respectively. These heat rates increase about 15–20% over the course of the following 9 h heating operation. These analyses do not consider the snow melting process but rather a constant ambient temperature. In any case, these heating rates can be compared with the estimated heating requirements presented earlier. As an example, the estimated heat flux requirement for −2°C ambient air temperature for no wind condition is about 20 W/m^2 as shown in Figure 11.3. It can be concluded that these temperature levels for the inlet fluid are sufficient to provide enough heating power to keep the deck surface snow/ice free for the 1 cm/h snowfall considered for estimating the heat flux requirements.

The effect of circulation tube spacing is also investigated, as this could be one of the primary design factors. The heating durations required to reach 0°C average deck surface temperature are shown in Figure 11.8 for different tube spacings. In this case, the concrete cover thickness above the tubes is 5.7 cm and the ambient air temperature is −2°C. Tube spacing of 15 cm results in 19–30% reduction in heating time to reach 0°C the average deck surface temperature in comparison to 20 cm tube spacing for different inlet fluid temperature. Similarly, 25 cm and 30 cm tube spacings result in 20–38% and 41–85% increase in heating periods, respectively, in comparison with 20 cm tube spacing. Apparently, it is possible to heat the bridge deck using these tube spacings within reasonably short time periods for the range of considered inlet fluid temperatures.

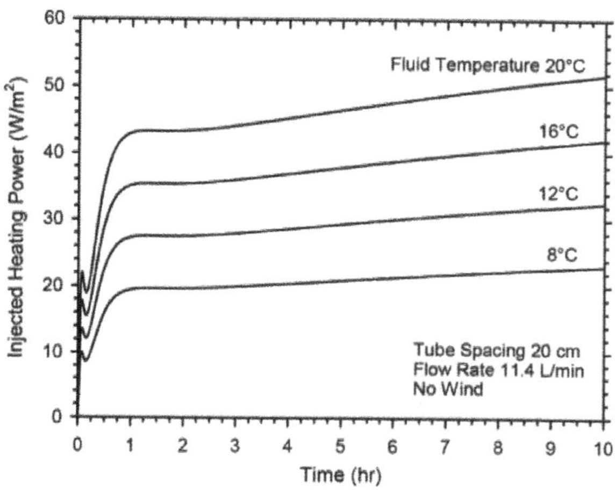

Figure 11.7. *Injected heating power for unit deck surface area*

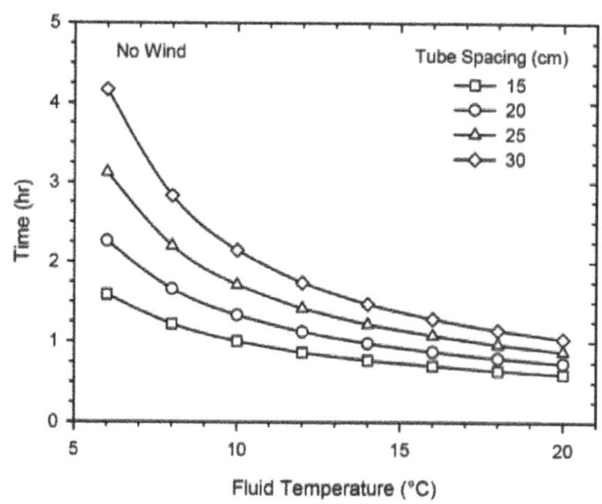

Figure 11.8. *Heating time required to raise the average deck surface temperature to 0°C for different circulation tube spacings*

The results of the analysis related to the effect of wind speed are summarized in Figure 11.9, where the heating times required to reach 0°C average deck surface temperature are shown for different wind speeds. It is seen that wind has a significant effect on deck heating for lower inlet fluid temperature.

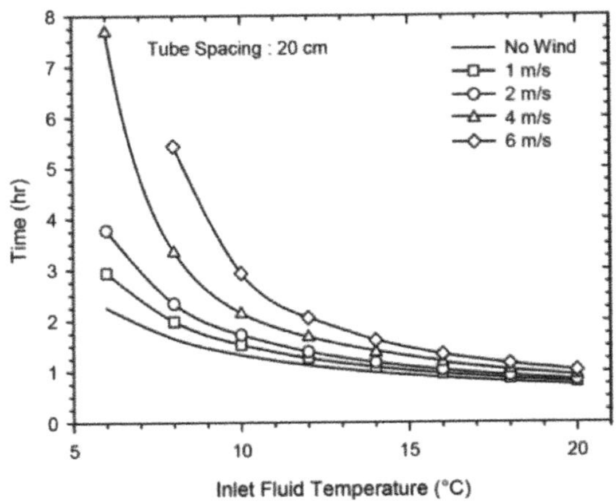

Figure 11.9. *Heating time required to raise the average deck surface temperature to 0°C for different wind speed*

Thickness of the concrete overlay above the circulation tube can vary due to a variety of project considerations. Therefore, our analyses also included cases where the thickness of the concrete cover was varied from the base case value of 5.7 cm. The results for the heating time to reach 0°C average deck surface temperature are shown in Figure 11.10 for different concrete cover thicknesses. It appears that every centimeter of cover thickness adds about 10–20 min to the bridge deck heating for the above metric. Within this given range, the thickness of the concrete cover has a smaller effect on heating time for closer spaced tubes in comparison to more distantly spaced ones.

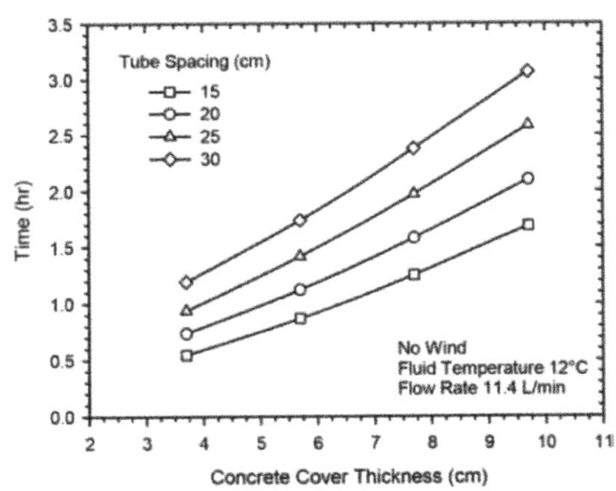

Figure 11.10. *Heating time required to raise the average deck surface temperature to 0°C for different concrete cover thicknesses*

The results for the effect of the fluid flow rate on deck heating are not presented here. Several analyses were performed for 6, 8, 15 and 20 L/min flow rates in comparison to the 11.4 L/min in the base case. The effect of flow rate is relatively minor and slower flow rates result in longer times to heat the bridge deck. A flow rate of 6 L/min results in a 13% increase in heating time whereas 20 L/min results in a 6% decrease in the heating time compared to 11.4 L/min flow rate. Analyses were also performed to investigate the effect of ambient temperature on the bridge deck heating performance. It was observed that every °C reduction in ambient temperature resulted in a 40–50% increase in the time required to raise the average deck surface temperature to 0 °C.

11.5. Summary and conclusions

Ground-source heating of bridge decks can be an alternative to the use of salts and chemicals to de-ice bridge decks. Energy foundations, geothermal boreholes, shallow trenches or the approach embankment can be utilized as energy storage media and heat exchange pathways to utilize the ground as a heat source. This can be a cost effective and sustainable method in comparison to the potentially detrimental effects of salts and chemicals. Warm fluid can be extracted from the ground and circulated through the tubing embedded in the bridge deck slab. Fluid circulation can be performed using either a heat pump or a circulation pump. Heat pumps (compared to circulation pumps) require more electrical energy and, therefore, necessitate the use of an electrical power source for operation. While the heat pump can provide higher inlet fluid temperature and result in more effective bridge heating, circulation pumps rely more directly on the *in situ* ground temperature. These distinctions separate the use of heat pumps and circulations pumps for different types of de-icing applications, i.e. heat pumps are more suitable for larger scale projects with high heating demands, and circulation pumps are suitable for smaller size applications.

This chapter has outlined the operational principles and how these are related to the design parameters of bridge deck de-icing systems. A series of parametric analyses were performed to investigate the bridge deck de-icing process. The analyses considered a variety of tube spacings, inlet fluid temperatures, flow rates, wind speeds, ambient temperatures and thicknesses of concrete cover over the circulation tubes. The results serve as a benchmark to gauge the operational conditions and the energy requirements for designing ground-source bridge deck de-icing systems.

11.6. Acknowledgments

This chapter is based on the work supported by the National Science Foundation under grant CMMI-0928807. The second author is supported by the National

Science Foundation as a graduate research fellow. Both supporters are greatly appreciated. Any opinions, conclusions or recommendations expressed herein are those of the authors and do not necessarily reflect the views of the National Science Foundation.

11.7. Bibliography

[ACI 11] ACI COMMITTEE 318., *Building Code Requirements for Structural Concrete (ACI 318-11) and Commentary*, American Concrete Institute, Farmington Hills, MI, 2011.

[AME 11] AMERICAN SOCIETY OF HEATING REFRIGERATING AND AIR-CONDITIONING ENGINEERS (ASHRAE)., *2011 ASHRAE Handbook – Heating, Ventilating, and Air-Conditioning Applications (SI ed.)*, American Society of Heating, Refrigerating and Air-Conditioning Engineers, Inc., 2011.

[BAB 91] BABOIAN R., "Synergistic effects of acid deposition and road salts on corrosion", *Symposium on Corrosion Forms and Control for Infrastructure*, San Diego, pp. 17–29, 3–4 November 1991.

[BER 11] BERGMAN T.L., INCROPERA F.P., DEWITT D.P., *et al.*, *Introduction to Heat Transfer*, 6th ed., Wiley, Hoboken, 2011.

[CAD 83] CADY P.D., WEYERS R.E., "Chloride penetration and the deterioration of concrete bridge decks", *Cement, Concrete & Aggregate*, vol. 5, no. 2, pp. 81–87, 1983.

[FED 08] FEDERAL HIGHWAY ADMINISTRATION (FHWA)., Status of the nation's highways, bridges, and transit: conditions & performance, Report to Congress, U.S. Department of Transportation Federal Highway Administration, Federal Transit Administration, Washington D.C., 2008.

[GRA 09] GRANATA R., HARTT W., *Integrity of Infrastructure Materials and Structures*, FHWA-HRT-09-044, Federal Highway Administration, Washington, D.C., 2009.

[HAV 78] HAVENS J., AZEVEDO W., RAHAL A., *et al.*, "Heating bridge decks by electrical resistance", *Proceedings of the 2nd International Symposium on Snow Removal and Ice Control Research*, Special Report 185, Hanover, pp. 159–168, 15–19 May 1978.

[HOP 01] HOPPE E.J., Evaluation of Virginia's first heated bridge, Transportation Research Record, No. 1741, pp. 199–206, 2001.

[IMB 85] IMBSEN R.A., Thermal effects in concrete bridge superstructures, NCHRP Report No. 276, TRB, National Research Council, 1985.

[KAY 80] KAYANE I., "Groundwater use for snow melting on roads", *GeoJournal*, vol. 4, no. 2, pp. 173–181, 1980.

[KOC 02] KOCH G., BRONGERS P., THOMPSON N., *et al.*, Corrosion costs and prevention strategies in the United States, Report No. FHWA-RD-01/156, Federal Highway Administration, Washington, D.C., 2002.

[KUM 88] KUMAGAI M., NOHARA I., Studies on a practical use of snow melting system by using the heat of ground water through pipes, Report No. 41, National Research Center for Disaster Prevention, Japan, pp. 285–309, 1988.

[LIU 07] LIU X., REES S.J., SPITLER J.D., "Modeling snow melting on heated pavement surfaces. Part II: experimental validation", *Applied Thermal Engineering*, vol. 27, no. 5–6, pp. 1125–1131, 2007.

[MIN 99] MINSK L.D., Heated bridge technology: report on ISTEA Sec. 6005 program, Publication FHWA-RD-99-158, FHWA, U.S. Department of Transportation, 1999.

[NAI 10] NAITO C., SAUSE R., HODGSON I., *et al.*, "Forensic examination of a non-composite adjacent precast prestressed concrete box beam bridge", *Journal of Bridge Engineering*, vol. 15, no. 4, pp. 408–418, 2010.

[POU 04] POUPARD O., AIT-MOKHTAR A., DUMARGUE P., "Corrosion by chlorides in reinforced concrete: determination of chloride concentration threshold by impedance spectroscopy", *Cement and Concrete Research*, vol. 34, no. 6, pp. 991–1000, 2004.

[SPI 00] SPITLER J.D., RAMAMOORTHY M., "Bridge deck deicing using geothermal heat pumps", *Proceedings of the 4th International Heat Pumps in Cold Climates Conference*, Aylmer, Québec, see http://www.hvac.okstate.edu/research/Documents/HPCC_GLHEPRO.pdf, 17–18 August 2000.

[TUA 04] TUAN C.Y., YEHIA S.A., "Evaluation of electrically conductive concrete containing carbon products for deicing", *ACI Materials Journal*, vol. 101, no. 4, pp. 287–293, 2004.

[TUA 08] TUAN C.Y., "Implementation of conductive concrete for deicing (Roca Bridge)", Publication SPR-P1(04) P565, Nebreska Department of Roads, Materials, and Research, 2008.

[VIR 83] VIRMANI Y., CLEAR K., PASKO T., Time-to-corrosion of reinforcing steel in concrete: vol. 5 calcium nitrite admixture or epoxy-coated reinforcing bars as corrosion protection systems, Report No. FHWA-RD-83/012, Federal Highway Administration, Washington, D.C., 1983.

[WAN 06] WANG K., NELSEN D., NIXON W., "Damaging effects of deicing chemicals on concrete materials", *Cement & Concrete Composites*, vol. 28, pp. 173–188, 2006.

[YEH 00] YEHIA S.A., TUAN C.Y., "Thin conductive concrete overlay for bridge deck deicing and anti-icing", *Transportation Research Record*, no. 1698, pp. 45–53, 2000.

[YUN 03] YUNOVICH M., THOMPSON N., VIRMANI Y., "Life cycle cost analysis for reinforced concrete bridge decks", Paper No. 03309, *CORROSION/03*, San Diego, CA, 2003.

[ZHA 10] ZHAO H., WANG S., WU Z., *et al.*, "Concrete slab installed with carbon fiber heating wire for bridge deck deicing", *Journal of Transportation Engineering*, vol. 136, no. 6, pp. 500–509, 2010.

PART 3

Engineering Practice

Chapter 12

Delivery of Energy Geostructures

Aspects that need to be considered to ensure the successful planning, design, construction and commissioning of energy geostructures are discussed in this chapter. The focus of the chapter is on issues associated with these systems that might not be immediately obvious to those new to the field. This guidance has been supplemented with a number of case study examples sourced from various countries and using differing types of structure for thermal exchange.

12.1. Introduction

To deliver an energy geostructure that performs satisfactorily, the key issues to be dealt with are 1) ensuring that the system is installed without major loss of heat exchange (HX) potential through component damage and 2) the installation process associated with the HX system is merged as seamlessly as possible into the construction process, avoiding undue delay and additional cost.

Each new energy geostructure has an associated learning curve for all stakeholders involved in the project, and there is a need for each to carefully assess the impact that the integration of such structures has on their interests, in terms of delivery of the wider project. Consideration of interfaces during design and construction, additional measures and/or processes in the work stream and many other factors are also required. These considerations are discussed in the following sections.

Chapter written by Peter BOURNE-WEBB with contributions from Tony AMIS, Jean-Baptiste BERNARD, Wolf FRIEDEMANN, Nico VON DER HUDE, Norbert PRALLE, Veli Matti UOTINEN and Bernhard WIDERIN.

The case studies presented within this chapter are drawn largely from projects in countries where the use of energy geostructures is fairly mature, that is Austria, Germany and the United Kingdom. However, even in these countries, the learning experience continues as energy harvesting is applied to new types of structures, for example tunnel linings, and under differing conditions. In addition, case studies representing the first applications of energy foundations in Finland and France are presented.

12.2. Planning and design

12.2.1. *Coordination and communication*

In a project that plans to use energy geostructures, it is important that there is a clearly defined set of responsibilities for each stakeholder involved from the outset, and that the interactions between stakeholders are managed by a contractually defined coordinator in the preconstruction phase, during construction, and even in the postconstruction period.

Where the responsibility for this role will lie may depend on the management of the project and on the contractual arrangements that are in place. There are many possibilities, but ultimately the choice should probably reflect where the best overview of the various stakeholders' needs is held:

– All designs of (geo) structures and systems by engineering professionals with contractors building the elements to design specification requirements. Here, the coordinator should perhaps be from the design team but will also need a strong rapport with the construction team, to ensure all the requirements for the project are communicated to the appropriate construction professionals and trades.

– Design of ground source heat pump (GSHP) system by a specialist contractor working with the engineering professionals forming the design team who is then novated to the construction team. In this case, the coordination role might fall to the specialist contractor, perhaps with a representative seconded/remaining with the design team.

– Design of most of the project elements (geostructure and GSHP system, etc.) by a specialist subcontractor to the main contractor. Here, coordination might best be the responsibility of the main contractor who will need to maintain a good working relationship with both the client's design team and his/her subcontractors to ensure the ease of information flow back and forth between the various parties involved. Guidance prepared by the Ground Source Heat Pump Association in the United Kingdom [GRO 12] recommends this approach for the construction and fit-out stages of a project; however, it is important that the design principles for the energy system, established before construction, are upheld within the subcontracted design process.

Case study CS12.1 illustrates how this process is being managed in one of France's first energy foundation projects; interestingly, the geothermal designer takes responsibility for assessing all the thermal effects in the piles (energy supply and pile–soil interaction effects) and the piling contractor dimensions the piles appropriately based on the effects defined by the geothermal designer.

12.2.2. *Design management*

Although it is possible, it is unlikely that all the different elements of an energy geostructure will be designed and detailed within the same organization; it is more likely that some or all of the various elements will be designed by specialists from different organizations (either engineering consultants or contractors). Therefore, the coordination role described in section 12.2.1 will need to be applied in the preconstruction phase. Hence, to ensure that project overview is maintained and to reduce the risk of underperformance of the energy geostructure, the same coordinator might be used throughout the differing project phases from conception to commissioning.

Design stage interfaces are very important for the successful delivery of a project that incorporates an energy geostructure; Table 12.1 summarizes some of these interactions from the perspective of the differing specialists who will be involved. The detailed requirements for design in each of these specialties can be found elsewhere (e.g. [GRO 12, Chapter 13]).

12.2.3. *System design redundancy*

In addition to the need for clear communication between the various design professionals, at certain points in the preconstruction phase, reviews of the proposed requirements of the system should be provided. These reviews should ensure that all design changes have been incorporated and that there is a sufficient redundancy between the design outputs of the system and that which is intended to be provided, to account for system underperformance, changes in expected ground conditions, and damage or loss of HX circuits.

A default figure of 10% would be a typical starting point; that is the GSHP system would be designed on the basis of an energy output 10% greater than the design value defined by the building needs. This margin (in effect a factor of safety on the energy performance) might however be increased or decreased depending on the confidence that the stakeholders have in the execution of the project – their previous experience in construction of these systems, performance of similar systems on similar ground and with similar operational conditions, the potential for changes to occur during construction and so on.

Case Study 12.1. *Business incubator in Bayonne, ECOME/ETCHART*

Introduction

In the delivery of ground source energy systems, the philosophy adopted by ECOME-Géothermie-Professionnelle of France is a systems approach, which recognizes that in order to deliver an energy geosystem successfully, conception, design and construction cannot be separated. Furthermore, geothermal pile design is a multidisciplinary task involving specialists in thermal, hydraulic, geotechnical and structural designs. Thus, to maximize the chances of a successful outcome, continuity between these disciplines is of the utmost importance to ensure the interactions between all the elements are correctly taken into account.

The building company ETCHART competed in a design and build competition for a new business center in Bayonne (Southwest France; artists impression above) for which a high level of environmental efficiency was required. The following case study demonstrates the approach used by ECOME in providing a geothermal pile solution that met these requirements.

Tender competition conceptual design

ETCHART commissioned the specialist thermal consultant ECOME to provide a feasibility assessment and cost estimates regarding the use of the foundation piles to cover the heating and cooling needs – a total of 60 kW heating and cooling was needed.

The thermal assessment was made using a simplified numerical simulation with the soil's thermal conductivity estimated from the published data. Working with the heating ventilation and air conditioning (HVAC) designer, ECOME optimized the building energy system, including the use of 64, 12 m long continuous flight auger (CFA) type bearing piles to provide 100% of the required heating and cooling. ETCHART decided to include the geothermal pile system in their proposition, including heating and free cooling, and subsequently won the tender for design and construction.

Execution stage design

Following the award of the contract, an *in situ* thermal response test (TRT) was undertaken and, working with their partner company, Enercret GmbH, a detailed numerical simulation of the ground energy system was undertaken.

ECOME has cooperated in the development of a design process and tools that take into account the mechanical impact of the thermal variations in the design of geothermal piles; a process that is certified by the Centre for Building Science & Technology (CSTB) in France. It addresses the following aspects of response:

– internal stresses and required resistance of the pile;

– pile–soil interaction and mobilization of soil resistance;

– impact on the structure supported by the geothermal piles due to differential settlement.

The contract pile design resulted in the foundation piles being slightly shorter than estimated in the tender design. This initial design was passed to ECOME to evaluate the energy system and the thermal effects on the pile elements. Due to the shorter piles used, in order to meet the design energy demand, the number of thermally activated piles was increased to 70. The results of the thermomechanical analysis of the pile foundation were reported back to the piling contractor and as a result, the diameter of three piles was increased in order to reduce predicted internal stresses to an acceptable level.

ECOME optimized the layout of the HX circuit based on the building energy supply needs and the thermal interaction analysis discussed earlier. Hydraulic data were then passed to the HVAC designer to confirm the circulation pumping requirements. Finally, a detailed methodology was defined in order to check the integrity of the system at the various stages of construction.

As already noted, good coordination between ECOME (geothermal company), the piling subcontractor, the general contractor and the HVAC company is of critical importance to ensure that responsibilities are clearly defined, and communication and implementation occur without problems, as shown in Figure 12.1.

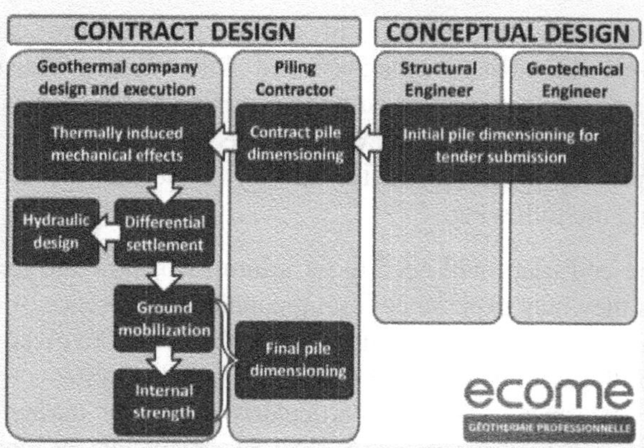

Figure 12.1. *Process map for mechanical impact of thermal effects*

Element	Responsibilities
M&E/ HVAC	Those responsible for the mechanical and electrical (M&E) or heating ventilation and air conditioning (HVAC) components of the system, that is the parts that deliver and/or extract heat from the structure needing heating and/or cooling, have to develop suitable models in order to predict the energy demands for the structure and thus define the operational requirements and control systems for the ground energy system, and any other heating/cooling elements that may be required.
GSHP	Those responsible for integrating a GSHP energy system into a building's climate control system will need to interact with the designers of both the M&E/HVAC systems and the geostructure in order to define what proportion of the heating/cooling load can be provided by the system. The GSHP designer needs to make the system design needs clear to all parties.
Structure	Those responsible for the design and provision of the geostructure need to be aware of the requirements for the integration of the heat exchange elements within the structure and the impact that changes to the dimensions of the structure have on heat exchange potential. It is vitally important that such changes (e.g. pile toe levels, pile cutoff levels) are communicated to the GSHP designer as the output of the system may be critically altered and other provisions need to be made to ensure the design thermal load is provided.

Table 12.1. *Design interfaces for energy geostructures*

12.2.4. *Awareness and skills training*

Education regarding the delivery of energy geostructures can probably be divided into two broad areas – awareness and skills training. Raising stakeholder awareness in terms of what is involved and what impacts the inclusion of an energy geostructure within a project might have, may need to occur at several levels within the scope of a project, depending on the novelty of the concept within a particular market, including the following:

– Clients and end users will need to be informed of how the technology works and the benefits they may experience from the inclusion/presence of a GSHP system in their structure. They also need to understand that those involved in the delivery and operation of the systems need to be aware of how the systems work and how to optimize their performance. This is particularly important during the life of the structure when owners/occupiers and facilities management companies might change.

– Regulatory bodies need to be contacted as early as possible in the planning and design process, and their requirements will vary significantly from country to country, and perhaps regionally. In any case, it is likely that there will be a need to familiarize the body/bodies with the concepts of the technologies, and to be able to deal with queries relating to many issues, such as impacts on groundwater flows and water quality. Starting from a position of complete unfamiliarity with these systems, this process could take some time and have implications for the timing and the delivery of the project.

– Neighbors and affected property owners may need to be made aware of how the technology works and its likely impacts, or rather be satisfied that heat pollution will not be a problem for neighboring sites. The need for permissions to exploit the ground energy potential below a property is highlighted by Frodl *et al.* [FRO 10] and in case study CS.6. In the future, property owners may seek compensation for the loss of utility to themselves implied by such systems, should the ground energy resource develop a significant value.

– The design team needs to be familiar with the technology and have the skills needed in order to deliver the project successfully – this is self-evident. GSHP systems require a much more refined understanding of the energy supply demand than conventional heating systems, and as systems get more complex, the need for more advanced types of analysis will arise – for example, detailed thermal simulations, maybe coupled with ground models to assess energy supply potential, impacts on groundwater and/or the extent of thermal plumes at property boundaries.

– The construction team needs to be aware of the technology and have the skills and experience of integrating these systems into the construction process – this may severely limit the number of groups that can tender for and undertake this type of project. Where these skills do not exist or are limited, there will be a learning curve for all involved and this needs to be recognized in the construction planning and budget. It is probable that many differing trade groups will be affected by the introduction of the HX system and it will require careful coordination and clear briefings to each group to ensure that the work is completely successfully.

– After construction, the GSHP system will be handed over to the building owner and the facilities management team that will operate the building systems. The latter will have to understand how the system works and how to integrate it (along with conventional systems) into the building climate control system in order to optimize performance and maintain the benefits to the current building owner/occupier.

In addition to simple awareness, the need for specific skills training might need to be considered within each of the design, construction and operational teams involved. Currently, this has to be dealt with within a project; however there is a

European Union-wide scheme (GEOTRAINET) for the training of operatives in the installation of borehole heat exchangers (BHXs). Perhaps this type of scheme should be expanded to include the particular requirements for energy geostructures.

12.3. Construction

12.3.1. *Process quality control*

In conjunction with the coordination aspects described earlier, a parallel process of monitoring and quality control needs to be implemented during the construction phase in order to ensure that damage or loss of function within the HX circuits is minimized. This needs to be managed by the main contractor with assistance from the geothermal system provider, and the requirements need to be clearly defined (and priced) in the construction contract.

Visual inspection of the HX pipes at various stages of the supply and installation process should be undertaken–that is the pipes should be checked for cuts and abrasions that might lead to leakage and kinks, which may restrict flow through the pipe circuit.

Pressure and flow testing should be mandatory at each stage of the work where there is a change in the work force operating on or in proximity to the HX pipe network. This might be before departure and after the delivery of prefabricated elements to site, during concreting of pile/wall elements (Figure 12.2), after completion of piling/walling works, before start of groundworks, and so on.

Suggestions for the timing of inspection/testing hold points and details of the types of test that should be undertaken for an energy foundation type system and recommendations with respect to the level of acceptable damage are provided in [GRO 12].

An interesting contractual approach was used to define responsibilities in the construction of works including HX elements for the Vienna metro [UNT 04]. It was accepted that there was a systemic potential for some damage/loss, however, it was also recognized that the contractor had a responsibility to limit problems beyond this background value. Thus, the contract was formulated such that loss of function in up to 3% of the total absorber length was considered unavoidable (this would also be included in the design margin between supply and demand, see section 12.2.3.), but beyond this point the contractor had to make good any loss and had to guarantee that the rate of failure would not exceed 12%.

Figure 12.2. *Pressure testing of HX loops on a diaphragm wall prior to panel concreting (left); HX loop pressurized during concreting and for 1 day afterward (right), with permission of GI Energy*

12.3.2. *Installation details*

12.3.2.1. *Pile foundations*

At the present time, a wide range of differing pile types – micro-piles, precast piles, driven cast *in situ* (DCIS), CFA and rotary bored piles – have been adapted to include HX loops.

The issues with each type are largely similar, i.e. how to fix the pipes, how to integrate the pipe work with minimal impact on the construction process and program, and how to prevent damage during installation and follow-on works, especially where piles need to be broken down. Many of the issues associated with particular pile types have been addressed, and these are discussed in the following.

Concrete trimming activities at the pile head provide the greatest potential for damage to HX pipes; however, robust and proven means exist for providing adequate protection. Typically, the inflow and outlet pipes will be gathered together, wrapped with either debonding foam (also used on the reinforcement steel at the head of the pile) or similar material, and run through a rigid plastic sleeve, as shown in Figure 12.3.

Figure 12.3. *HX pipe protection installed at pile head, prior to concreting, with permission of Cementation Skanska*

12.3.2.1.1. Precast concrete piles

Prefabricated under controlled conditions in a factory setting, precast concrete piles can be readily adapted for use in a GSHP system. However, the pipes require protection from damage during the installation process, which involves driving the pile by impact at the pile head. This is normally achieved either by having the HX pipes exit the pile body some distance below the head of the pile, or by breaking out of the pipe work from the pile body after installation.

Case study CS12.2 describes a project using precast concrete piles for an office development constructed in Austria in the early 1990s and how the same technology was used with some system improvements for an extension constructed in 2008.

The main commercial advantage of this type of pile is that the piles can be mass produced (low unit cost) and installed quickly. The work process cannot accommodate lengthy delays (e.g. for jointing of HX pipes) and thus, HX loops have typically been limited to the uppermost section of piles typically less than 15 m. However this is often sufficient for lighter types of construction, for example low rise residential. An economically viable jointing system that incorporates HX pipes and allows deeper precast pile installations has yet to be brought to market.

12.3.2.1.2. Hollow driven and screw piles

This type of pile may be made from steel, cast iron or spun precast concrete and installed by driving into the ground. Onshore, these types of pile tend to have diameters less than approximately 0.6 m and may use one or two HX loops.

The HX loops are lowered into the internal void of the pile that is then backfilled with grout. In Japan, alternative forms in steel use either water as the infill material surrounding the loops, or use an open circulation system within the tube (Figure 3.1, Chapter 3). In some types of tube pile, grout is injected as the pile is inserted into the ground and the HX loops need to be plunged into the liquid grout, which may limit the length of loop that can be installed.

Case study CS12.3 describes such a system used in Finland and a novel means for leading the pipes into the building, which took the installation of the pipes away from the groundwork stage, and off the build program's critical path.

12.3.2.1.3. Driven cast *in situ*

DCIS piles (also known as "Franki" piles) are formed by driving a closed-end steel tube into the ground, placing a reinforcement cage (if needed) inside the tube, charging the tube with concrete and then extracting the tube (leaving behind the end cap that closed the end of the steel former tube), topping up the concrete as necessary.

Often, the pile only carries a vertical load and, in the absence of a reinforcement cage, a DCIS pile will have a single central bar to provide continuity and protect against construction-related impacts, which might cause the pile to break. In this mode, it is possible to plunge the HX pipes having attached them to this central bar – the limits to this approach are discussed below in reference to CFA-type piles. Clearly, when a cage is required, the HX pipes can be attached to this before placing it in the pile bore and concreting.

Case Study 12.2. *Precast piles, ENERCRET – Nägelebau, Röthis, Austria*

Introduction

ENERCRET GmbH has specialized in the installation of energy piles and other energy geostructures since the early 1980s and when, in 1990, the new offices (pictured above) of their parent company Nägelebau GmbH was built in Röthis (Austria), and was founded on bearing piles, it was a natural outcome that the foundations should be thermally activated.

A total of 52, 12 m long precast piles were required and were thermally activated in order to provide heat to a footprint area of 934 m^2; heat is distributed in the building via an under-floor system.

Cooling by "free cooling"

Originally, the need for cooling was not considered in the design, but later as the need for cooling became apparent, a simple solution had to be found. "Free cooling" the piles in combination with the under-floor heating system seemed to be a simple and inexpensive solution. The term "free cooling" is used in the situation where HX is obtained from the circulation of the HX fluid by pumping, without the use of the heat pump, thus saving the cost associated with running the latter.

In this case, because of the limited pressure head in the system (less than 2 bar), even greater cooling efficiency could be achieved because the ground loop circuit could be connected directly to the building circuit, circumventing the heat exchanger interface. Flowing groundwater also contributes to the thermal exchange potential and the ground temperatures are never higher than 16°C.

Improvement

Because of the success of the energy foundation system, it was also used for an extension of the building constructed in 2008. An additional 49, 10 m long piles were thermally activated to provide heating and cooling to the extension. The center of this system is a 57 kW heat pump with two compressors.

As an improvement, the new part of the building uses air ventilation to guarantee fresh air in the offices. This air flow is preheated in winter and precooled in summer (the latter by "free cooling"). That way a part of the heating and cooling is distributed by air, whereas the main part is still provided by the underfloor system.

Thermal performance

Approximately 2,200 m² of the building is heated in winter and 1,970 m² of the building is cooled in summer. For heating, approximately 20,000 kWh of electricity is used per year (costing 2,400€). In heating mode, the coefficient of performance (COP) of the system including the circulation pump is approximately 4.3. The heating energy supplied by the system during 2011–2012 is illustrated in Figure 12.4.

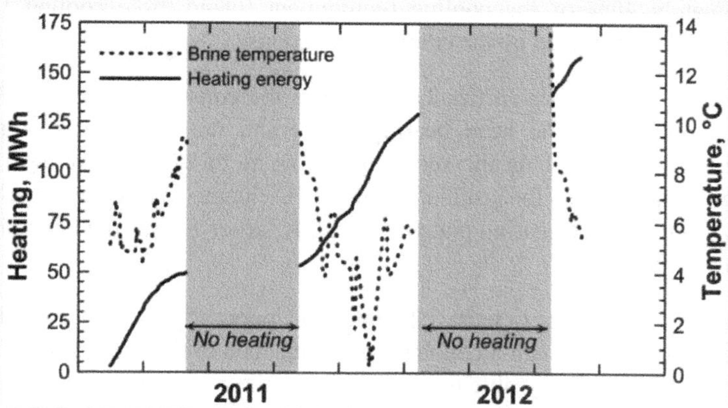

Figure 12.4. *Heating energy and associated circulation fluid (brine) temperatures during two years of operation of the energy foundation*

For cooling the building, only a 450 W circulation pump is used. The electrical energy for the pump in the cooling period is 620 kWh per year, costing just 74€. The capacity of this cooling system is approximately 35 kW and the room temperature is maintained below 26°C in summer.

Case Study 12.3. *Driven steel tube piles, Ruukki, Jyväskylä, Finland*

Introduction

The first application of energy foundations in Finland was for a six-floor office building in the town of Jyväskylä [UOT 12]. The site is situated on a reclaimed lake and is underlain by up to 29 m of fill, soft organic clay/silt and loose to dense silt. These lacustrine deposits are followed by a layer of glacial till and bedrock. The groundwater table is 1–2 m below the soil surface.

A total of 246 Ruukki RR-type, driven steel pipe pipes, with diameters of 170 mm and 220 mm and lengths between 22 m and 29 m were used to support building service loads of 691 kN to 1350 kN. Groups of three to four piles were used to support the column loads and a number of piles were also used to support the base slab. The piles were installed using a 4T hydraulic hammer and pipe sections were joined by internal friction sleeves in the larger diameter piles and external sleeves in the smaller diameter piles.

Energy pile system and configuration

One pile from each group (32) and six base slab piles were modified for use as energy piles, an overall spacing between energy piles of 5.5–7.8 m was maintained. In this configuration, it is estimated that about half the heating and approximately 40% of the cooling requirement (using "free cooling") of the building will be provided by the energy foundations.

The only change of significance needed in the construction process was the modification of the pile head bearing plates and the pile cap reinforcement (Figure 12.5, top left). This allowed the pile body to be extended through the pile cap and provide access for installation of the HX loops (Figure 12.5, middle and right). As a result, the bearing plates are slightly larger and need a penetration for the tubes to pass.

Each set of HX pipes (a pair of 25 mm OD Uponor PE-Xa pipes) was lowered into the steel pipe pile with a counter-weight attached to the end. In two piles, the full depth of tube was not able to be installed as the pipes jammed in the internal sleeve at the joint between pile sections. After the insertion of the pipes, the remaining void within the steel pipe piles was filled with grout. For future projects, to avoid the problem described above, Ruukki now supply an external connection sleeve for larger diameter pipes to replace the internal sleeve fitting.

From the pile, the pipes were run to manifolds within a void space between the base slab and the actual floor slab, which is shared with other service ducts (Figure 12.5, bottom left). The pipes are insulated to prevent thermal loss. Insulated mains then connect from the manifolds to the plant room and the heat pump.

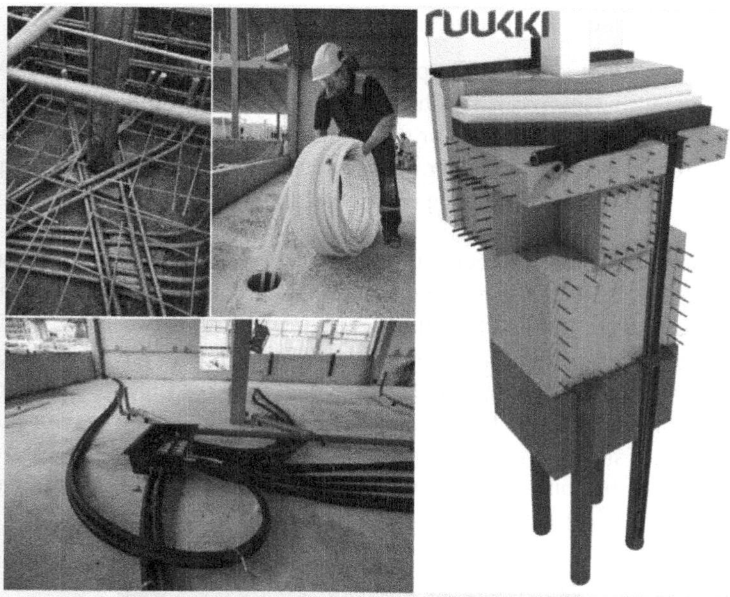

Figure 12.5. *Pipe pile concept (right); energy pile extended through pile cap of four piles (top left); HX pipes, one loop per pile, are inserted after groundworks and floor slab are finished (center top), followed by laying of pipe work back to control room (center bottom*

Summary

The solution based around driven pipe piles presented here has a significant advantage in that the HX loops could be installed after the works associated with construction of the pile caps and the building substructure (beams and base slab; see schematic opposite). This minimized the risk of damage to the HX pipes and reduced the number of construction interfaces to a minimum.

The only additional cost in the foundation construction was associated with the prefabrication and supply of the 38 modified bearing plates for the energy piles, and the sections of pipe used to extend the pile through the substructure to allow access from the base slab. The actual process of pile installation and the time for construction of the foundations was unaffected, as the HX pipes could be installed later in the building program, off of the critical path.

To conclude, the energy foundation system described has been operational since early 2012 and based on the energy production simulations, capital investment costs and the current price of energy, the repayment period for this scheme is expected to be approximately 9–10 years [UOT 12].

12.3.2.1.4. Continuous flight auger

Continuous flight auger (CFA) piles, (also called auger grouted piles (AGP)) are constructed by the insertion of a hollow-stem helical auger to the required depth followed by injection of concrete or grout through the stem as the auger and soil are extracted from the pile bore. Auger displacement piling is a similar technique that is increasingly used and the following discussion applies equally to this type of pile.

Because of the method of construction, it is necessary to plunge elements such as reinforcing cages and HX pipes into the still fluid concrete. The depth that a steel cage can be inserted in the pile is limited due to friction between the fluid concrete and the cage, the potential for concrete deeper in the pile to have hardened, and the weight and rigidity of the cage itself. Single or paired HX pipe loops are introduced into the pile while attached to a central bar by plunging, which may or may not be recoverable, as shown in Figure 12.6. The same rules of thumb used for cage installation must be considered when plunging HX pipe loops into freshly concreted CFA/AGP and DCIS piles.

Figure 12.6. *HX pipe loop attached to bar prior to plunging into CFA pile, with permission of Cementation Skanska*

Typically, a lightweight reinforcement cage might only enter to a depth of up to 8 m, whereas a heavy cage could achieve up to 15–18 m depth. Greater depths are possible, however the likelihood of the cage stopping short of the target depth increases dramatically. For HX pipe loops, depths of 20–25 m can be achieved, however, this is highly dependent on the particular site conditions and should be reviewed on a case-by-case basis. Additionally, the use of mechanical force or surging to increase the depth of penetration should be avoided due to the risk of damage to the HX pipes.

The risks associated with plunging the HX pipes need to be acknowledged (i.e. the required penetration depth not being achieved and the buckling of pipes) and taken into account in terms of the level of redundancy with respect to the design energy output required for the system, as discussed in section 12.2.3.

12.3.2.1.5. Rotary bored – dry

Rotary bored piles are constructed by exaction of the ground using augers or buckets depending on the ground conditions. Where the geological conditions permit, it is often possible to complete a rotary pile bore in the dry; this may entail the use of a section of recoverable casing to seal off near-surface water bearing soils.

Where the reinforcing cage is required to extend the full depth of the pile bore, the inclusion of HX pipe tubes is relatively straightforward as they can be fastened to the cage. Special consideration will need to be given to pipe continuity in the situation where the cage is spliced in sections. In general, it is preferred that pipe joints are formed offsite to ensure quality and to limit program delays, as forming joints is a time-consuming process. Also, there is always a risk that material will enter and block the pipes when forming joints in a field environment.

In the case of spliced reinforcement cages, one approach that has been used successfully was to prefabricate the HX pipe loops in their entirety and then attach the upper part to the upper cage section and fold the remaining loop length inside, for transportation to site. Then, once the prefabricated cage sections are on-site, the lower sections of the cage are spliced to the upper part successively and the HX loops are pulled down through, and fastened to each section as it is completed. Care must be exercised to ensure that none of the tubes are kinked.

Another situation that can arise is when the reinforcement cage needed for structural purposes extends part way down the pile shaft. Of course, the HX pipe installation could simply be limited to the depth of the cage; however, it is possible to achieve greater depths by either forming a light skeleton cage to support the pipe work or by suspending the pipes from the cage; here, the HX pipes can be held in position near the pile perimeter with the use of small sections of cage, so-called “lantern cages”, as shown in Figure 12.7.

Figure 12.7. *Lantern cage used for HX loop placement where cage is not of full depth, with permission of Cementation Skanska*

12.3.2.1.6. Rotary bored – wet

Often, rotary piles cannot be completed with a dry bore and need to be flooded with a fluid such as water, bentonite mud or polymer to ensure the stability of the sides of the bore and to balance water pressures.

As with piles that are dry bored, installation of HX pipe loops on full depth reinforcement cages is relatively straightforward, and again if the cage is not full depth, then a skeleton cage could be used to support the HX pipework below the level of the structural steel. While the use of a skeleton cage may be relatively expensive, the use of lantern cages is not recommended because in a fluid supported bore, it is not possible to visually verify that the pipes are hanging correctly and will not be caught by the concrete tremie tube, i.e. the risk of the tremie snagging and damaging the HX pipes is higher.

Spliced reinforcement cages present another challenge. In Frankfurt, large-diameter piles are constructed using water as a support fluid. Practice here has been to site weld the HX pipes across each section of cage, a process which is very time consuming. However, a large diameter pile could have 6 or 8 loops on the cage and the fusion welding process would add about one day per set of joints to the construction cycle time for each pile. Joints in the HX pipes also increase energy losses around the HX circuit, increasing the pumping effort required and hence, running costs.

In the United Kingdom, this approach is unacceptable both in terms of construction cost and delay, and technically, the longer the sides of the pile bore are left exposed to the support fluid, the greater the softening and lower the strength of the completed pile shaft. To enable the HX pipes to be installed with as little delay as possible and with as few joints as possible, the first thermally activated wet-rotary bored piles in the United Kingdom were constructed with the HX pipes moved from inside the reinforcement cage to the outside over most of the pile length. In the uppermost cage section, the HX pipes were passed back inside the cage in order to restore the concrete cover to the reinforcement and be safely led into the pile cap, as shown in Figure 12.8.

Figure 12.8. *Construction of bentonite supported rotary bored pile with HX pipe outside reinforcement cage, with permission of Cementation Skanska*

The potential for problems associated with the reduced cover to the reinforcing steel was considered to be minimal for three reasons: the HX pipes are attached mid-way between the longitudinal bars; below a few meters depth corrosion potential is minimal due to the lack of oxygen; near the surface where the corrosion potential is higher, the HX pipe is returned inside the cage, restoring the full cover. Should the piling environment be particularly aggressive, then one means for maintaining the concrete cover is to increase the pile diameter (or reduce the cage diameter); case study CS12.4 describes how this was done for a diaphragm wall.

12.3.2.1.7. Small-diameter piles

Micropiles (diameters up to 300 mm) and minipiles (diameters up to 600 mm) can be constructed using most of the techniques described earlier, and can also be adapted to include HX pipe loops. Often this type of piling is undertaken within existing structures and in confined spaces, therefore particular consideration and care is needed to ensure that the HX pipes are protected throughout the construction works.

12.3.2.2. *Embedded retaining walls*

Embedded retaining walls are often needed to support the sides of excavations for basements and underground transportation projects, whether in open cutting or cut-and-cover type tunnels, and may be formed using either piling techniques or diaphragm walls. The issues associated with the integration of HX pipe loops that need to be considered for bearing piles also impact the construction of thermally activated walls, that is protection of the tubes during installation and in the midst of follow-on activities.

There are, however, a number of additional issues that need consideration and case study CS12.4 describes how an overly complicated proposed HX pipe arrangement and the need for pipe continuity were addressed, without impacting the construction time, when delivering the first thermally activated diaphragm wall in the United Kingdom.

Typically, the HX pipes are routed from the wall elements via either the capping beam or at base slab level, that is at the lowermost floor level in the basement. In the former situation, the requirements are relatively straightforward, although allowance for the space taken by the pipes must be made within the details for the capping beam (Figure 12.9).

Figure 12.9. *HX piping incorporated into wall capping beam, with permission of GI Energy*

Case Study 12.4. *Diaphragm wall, GI Energy/Cementation Skanska, London*

Introduction

Completed during 2010, the Bulgari Hotel is the first example of a thermally activated diaphragm wall in the United Kingdom. The 800 mm thick wall with a panel depth of up to 36 m is required to support the excavation for a six level, 24 m deep basement and also vertical load from the superstructure. The wall extends through about 14 m of superficial soil deposits (fill, silty clay loam and river terrace deposits), into the London Clay Formation, which extends to depths below the basement structure. Groundwater stands about 10 m bgl, within the river terrace deposits.

To deliver 150 kW of peak heating and cooling, the ground energy system utilizes a 100 m plan length of wall (total 3600 m2), or about two-thirds of the total boundary and 49,25 m long thermally activated plunge column piles. It was not possible to thermally activate the entire perimeter wall because an existing deep basement was present along one boundary.

Considerations

The key issues in developing the thermally activated diaphragm wall solution concerned maximizing potential heating and cooling, in particular the impact of having an HX pipe on only one face of the wall elements, and minimizing construction complexity (and hence delivery risk) to ensure a near-continuous installation process and to maximize wall panel quality [AMI 10].

Thermal properties of the thermally activated wall

In this application, only one side of the wall is used for HX; this differs markedly from bearing piles where the entire element is surrounded by the ground. Thus, there was a great deal of uncertainty as to how this would affect the HX characteristics of the system. In the absence of supporting values, very conservative values for the thermal design parameters were assumed and a set of TRT planned in order to verify these assumptions. An initial TRT, undertaken prior to basement excavation confirmed the design values and a second TRT, carried out after excavation was completed, showed a reduction of 10% in conductivity value. This reduction was considerably less than expected, thus providing confidence in the values used for the scheme design.

HX pipe layout and installation

As the project moved to construction, the arrangement of HX pipes that was initially proposed by the consultant was reviewed and raised concerns that the complex pipe geometry, the details for attaching the pipes to the main reinforcement cage and the connection details where the pipes exit the wall panels, would result in major program conflicts, increased reinforcement congestion and could compromise the successful delivery of the system.

Based on their extensive experience with the construction of energy foundations, GI Energy Ltd and Cementation Skanska Ltd proposed a simpler arrangement of HX pipes based on the simple U-tube arrangement used in piles rather than the complex "slinky" form proposed [AMI 10]. Thus, within each panel, two loops were attached to the reinforcement cage on the side facing the surrounding soil mass; the pipes in each loop were spaced 0.5 m apart and the spacing between each loop was about 2 m.

Adopting a similar approach to that which had been used for deep rotary bored, bentonite supported bearing piles, the HX pipe loops were kept on reels next to the panel (Figure 12.10, left) and were attached to the outside of the cage as it was lowered into the completed excavation (Figure 12.10, right). Having the loops in the protective cover zone is not necessarily a problem for steel longevity; however, in this case, it was decided to provide concrete cover of 75 mm to the HX pipes thus increasing the effective reinforcement steel cover. This led to some alteration of the reinforcement layout but care was taken to ensure that congestion of the reinforcing steel did not worsen [AMI 10].

The process of HX loop attachment and pressure testing prior to concreting (Figure 12.2) did not delay the overall time for panel construction; this was controlled by the time it took to splice each cage section together, and pressure testing was undertaken while the piling contractor was placing tremie pipes for the concreting operation.

Figure 12.10. *Bulgari Hotel; HX pipe loops on reel (left) being attached to reinforcement cage in wall panel (right)*

Summary

This project exemplifies how good coordination at all stages, simple construction details, thorough preparation and a rigorous regime of pressure testing can successfully deliver a thermally activated wall system.

Case Study 12.5. *Penetrations and waterproofing, Bilfinger Berger, Germany*

Introduction

This case study is actually about three projects and illustrates the learning curve associated with the introduction of energy geostructures (in this case bearing piles) into the construction process [HUD 10]. The case study focuses on the sequence of adaptations made in the works by the contractor, Bilfinger Berger SE, to reduce complexity, to minimize HX pipe damage and to penetrate the building envelope without leakage.

A summary of each project is given in Table 12.2; large-diameter piles were bored using water as the support fluid to various depths up to 50 m bgl, HX loops were attached to prefabricated reinforcement cages and in the first two projects, the cages were delivered in sections and the loops were fusion-welded onsite during installation.

	Main tower	Gallileo	IG-Metall
Year completed	1999	2003	2004
Tower height (m)	195	136	90
Basement depth (m)	20	17.5	10
Pile diameter (m)	1.5	1.5	1.2
Pile length (m)	30	26–30	20
Number of piles	112	47	48
Number of loops per pile	8	8	6
Number of circuits per pile	8	2	1
Power rating (kW)	400	160	95

Table 12.2. *Summary of Frankfurt energy bearing pile projects*

Solutions

In each project, different arrangements for HX pipe circuits within the pile and pipe routes to the building climate control center were used as follows:

– Main tower: each pile had eight HX pipes entering and eight leaving (total 16) with a manifold collecting each of the eight pipes for the pipe-runs back to the control center. The pipes ran from each pile under the base slab to enter the building at several locations. The project suffered from a 10% failure rate in the completed HX circuits and leakage at the penetration points that required expensive mitigation.

– Gallileo: to address the problems experienced in the previous project, the loops inside the pile were connected in parallel to form two circuits and hence leave only two HX pipes entering and two leaving the pile. The four pipes leaving each pile were run under the base slab to a common point (Figure 12.11), fixed with a steel guide frame/spacer and led up through the slab.

To further limit the risk of leakage, two grout injection collars were provided on each pipe to allow for grouting later (Figure 12.11, lower right). In this case, only one pile circuit was lost and following the planned grout injection, the base slab did not leak.

– IG-Metall: in a further development, a new approach was used on this project whereby all the HX pipes within the pile were connected in parallel to leave only one inlet and one outlet pipe. These pipes were directly taken up, to the middle of the base slab where they were paired in parallel with an adjacent pile and then run within the slab back to a riser shaft, which then gave access to the building's interior and to the control center.

Figure 12.11. *Gallileo tower, Frankfurt: HX pipe runs from piles to location of base slab penetration (left), pipe frame at bottom of base slab (upper right), injection collars and grout lines in slab (lower right)*

Summary

In addition to illustrating the development of the solutions needed to address the particular issues discussed above, [HUD 10] also identifies a number of other considerations relating to the delivery of energy foundations, including:

– special requirements in the preparation of tender documents starting with the foundation works;

– coordination and cooperation in the design and construction phases;

– interfaces (between design professionals and construction trades);

– additional work steps within the construction process;

– quality assurance (control, pressure testing and documentation).

When the HX pipes exit the wall element at the base slab level, provisions needs to be made to protect the pipe ends during concreting and to ensure continued access when the pipe work at base slab level is being installed. This is achieved in a similar manner to reinforcement bars, using boxouts at the appropriate location (Figure 12.12) when fabricating the cage. Great care is needed to ensure that the boxout(s) do not affect the quality of the finished wall element, for example due to increased cage congestion, and that the boxout is correctly positioned before concreting.

Figure 12.12. *HX pipe exiting wall pipe from boxout (left) in order to feed into underside of thermally activated slab (right), with permission from Bilfinger Berger*

Often in embedded retaining wall construction, drilling and dowelling of starter bars is necessary to enable the construction of floor slabs and secondary linings. When this is proposed in conjunction with a thermally activated wall system, extreme care is needed to ensure that the drilling process does not puncture the HX pipes. The risk of damage is such that it may be prudent to consider alternative methods, i.e. providing starter bars within boxouts in the required locations.

12.3.2.3. *Headering-up, penetrations and waterproofing*

Having overcome the challenges associated with the integration of the HX pipes within elements such as piles or walls, detailing and planning must also consider where the pipes need to be routed in order to complete the thermal circuit with the heat pump. A key consideration here is where and how the pipework penetrates the building envelope and whether there is potential for leakage at this point; case study CS12.5 uses the example of three projects in Frankfurt to illustrate how the treatment of this issue evolved; pipe circuits were simplified and risk of leakage was minimized.

While the latter systems described in CS12.5 have prevented leakage into the basement, other systems have been employed successfully; e.g. puddle flanges and hydrophilic water-bar products (Figure 12.8) have been used in the United Kingdom.

Figure 12.13. *Waterproofing of a pipe penetration using hydrophilic strips, with permission of GI Energy Ltd.*

12.3.2.4. *Shallow foundations and rafts*

Horizontal elements such as ground bearing floor slabs and rafts (mats) can, of course, be thermally activated in the same manner as the vertical elements described in previous sections.

The issues of protecting the pipe work during installation and minimizing the impact on the construction program are the same and the need for close coordination of professions and trades remains. However, overcoming construction issues at a technical level is rather simpler than when dealing with vertical elements and is not discussed in detail here.

In many instances, HX pipes from pile/wall elements run either under or through slab/raft elements in contact with the ground (and groundwater), and by default these will be thermally activated (Figure 12.14). This should be incorporated in the thermal design assessment for the energy geostructure.

12.3.2.5. *Tunnel linings*

For many years, the use of warm water discharging from Alpine tunnels has been used for heating and cooling tunnel control centers, public facilities and residential properties in Switzerland [WIL 03]. Harnessing this beneficial by-product of tunnel construction makes sense. However, it will only be viable under certain circumstances and, in particular, by having a customer for the energy close to the tunnel portal.

Figure 12.14. *Distributor pipes to/from piles run within granular drainage blanket below base slab, with permission of GI Energy Ltd.*

Similar conditions are less likely to occur in urban environments, however potential customers are likely to be closer and the HX potential for underground spaces has been recognized for some time. A series of experimental/demonstration projects where tunnels have been used in ground source energy systems have been undertaken in Austria and Germany [BRA 06, FRO 10, SCH 10].

The first application of HX in a tunnel was the test section LT22 of the Lainzer Tunnel, Vienna, Austria where HX pipes were integrated into the construction of a sprayed concrete lined (SCL) or NATM-type tunnel [BRA 06].

SCL tunnel construction involves the mechanical excavation of the ground which needs to be either self-supporting or reinforced using soil nails, rock anchors, grouted pipe arches or other techniques, until a primary lining of sprayed reinforced concrete is applied. In a second operation, a secondary lining is laid sometime later. At LT22, in order to thermally activate this type of tunnel construction, the HX pipes were attached to a geofabric, which was then pinned inside the primary tunnel lining, before the construction of the secondary lining (Figure 12.15) [BRA 06].

A similar approach has also been used in the construction of two thermally activated test sections in the SCL Fasanenhof tunnel in Stuttgart [SCH 10]; here, however, the HX pipes are clipped into a set of plastic strips that are in turn fastened to the geofabric (Figure 12.16).

Figure 12.15. *HX pipes on geofabric attached to primary SCL tunnel lining test section at Lainzer tunnel, from [BRA 06]*

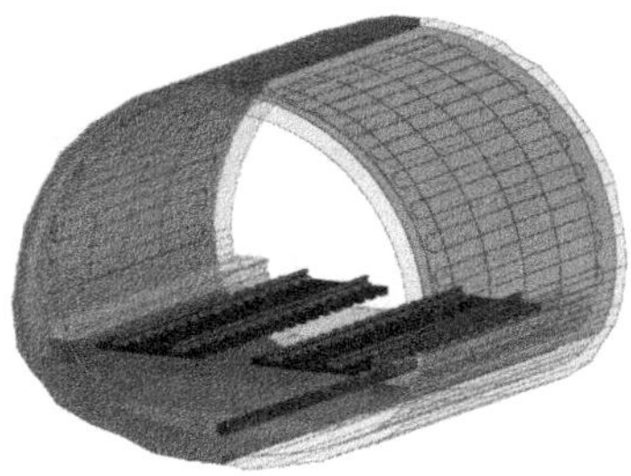

Figure 12.16. *Layout of HX loops and distributor pipes in thermal test section of Fasanenhof SCL tunnel, from [SCH 10]*

Another method of tunnel construction uses a tunnel boring machine (TBM) to excavate the ground, which is temporarily supported by a shield structure until precast segmental lining (SGL) elements can be placed, and the space between the tunnel lining and the surrounding ground can be grouted.

Case study CS12.6 describes the development and use of thermally activated tunnel lining segments for a demonstration project in Austria. A further alternative for application in tunnels could be to install HX pipes in the tunnel invert only [BRA 06]; in transportation tunnel construction, this zone is generally filled-in to provide the required pavement level, so there is no loss of utility to the tunnel operator.

Tunnel-based thermal exchange differs from that of foundations in a number of ways, as discussed in CS12.6 and later, and it is apparent that many of these differences relate to planning and preconstruction activity. The main technical difference is that thermal exchange may occur with the surrounding ground or the air volume within the tunnel. The air within near-surface tunnels (e.g. cut-and-cover types), short-run tunnels and near the portals of deep tunnels tends to be relatively cool, and thermal exchange with the ground will predominate.

Case Study 12.6. *Segmental lined tunnel, Ed Züblin, Jenbach, Austria*

Introduction

Many new tunnels are now being constructed using full-face TBM. This technique enables tunnel excavation in weak ground with a low risk of damage to adjacent structures. This type of tunnel construction uses precast concrete SGLs to support the ground.

This case study presents a system developed by Ed Züblin AG and Rehau AG that allows SGL tunnels to be turned into heat exchangers (SGL-HX) for the supply of heating and cooling to nearby buildings [FRA 11]. Located in Jenbach, Austria, the system was installed in a new 12 m diameter twin-track railway tunnel to trial the supply of heat to a municipal building. The tunnel runs through a geology dominated by saturated gravel layers at an invert depth of about 27 m. These conditions are ideal for this application, and SGL-HX units were installed over a length of 54 m (27 lining rings) to provide an estimated 15 kW of heating.

Considerations

The following considerations proved important in the delivery of the SGL-HX system at Jenbach [FRO 10]:

– regulatory controls,

– education of stakeholders,

– time needed to obtain permissions of affected landowners,

– delineation of the ownership of different parts of the system,

– control of access to the system for maintenance,

– the railway owner's requirement that future tunnel operations were not affected by the presence of the ground energy system.

In addition, the incorporation of an HX system into the SGL segments should meet the following requirements: structural integrity and water-tightness of the lining could not be compromised, fitting of heat exchange pipes should be integrated in the segment manufacturing process, pipe connections between adjacent segments could not interfere with the TBM advance, but should be robust enough to fulfill durability requirements and to cope with tunnel deformations.

Solutions

Energy lining segments developed by Ed Züblin and Rehau (Figure 12.17, bottom center) are fabricated in the same way as normal segments, the only difference being the absorber pipe work attached to the outside face of the reinforcement cage. Absorber pipe ends are exposed at boxouts on the inside faces of the segments (Figure 12.17, upper right), and the pipe fusion coupling system has been used successfully for many years [PRA 09].

To integrate the system into the tunnel without impact on the tunnel services and future operation, the pipe work associated with exchange fluid distribution was run along the invert of the tunnel to an adjacent escape shaft (Figure 12.17, upper left).

In the finished tunnel, the segments are covered by a secondary lining and the pipe work in the invert is covered by the railway track bed, thus pipe circuits will not be able to be accessed in the future. Therefore, the distribution controls for each circuit (tunnel ring) were housed in the escape shaft so as to allow access without interruption to the tunnel operation, see image below.

Summary

This tunnel-based HX system has been successfully operating since autumn 2012, and the energy yield has matched expectations. Further analysis of the data from the operational system will help us to optimize the operation of future systems.

The development and implementation of the SGL-HX unit has shown that TBM-driven tunnels can be used for economically viable heat exchange. The additional costs of integrating such systems are relatively minor compared with overall tunnel construction costs. The extra costs may not offer an investment return of less than 10 years; however, the lifespan of large infrastructure projects is such that this may not be relevant.

Energy lining segments for tunnels provide a new means for obtaining ground energy exchange within congested urban environments and could also be used in a number of other applications to replace energy intensive HVAC systems.

Figure 12.17. *Completed tunnel (upper left) with pipe runs to access shaft (mid-picture); schematic of precast tunnel segment (bottom) and tunnel lining segments in-place with fusion-welded pipe connections visible (upper right)*

However, in deep tunnels, the air within the tunnel tends to be warm (in the range of 20°C to 30°C) due to the build-up of latent heat from vehicle braking, air-conditioning, train idling, thermal losses from utilities, etc., and heat will be drawn from the ground and the tunnel air – the latter can provide some benefit to the tunnel owner in terms of cooling the tunnel (heat supply to end user). However, the injection of heat for cooling by the end user is unlikely to be welcomed by the tunnel operator.

The link between the thermally activated tunnel and the potential end user requires a great deal of consideration when tunnels are to be used for this application. Unlike foundations, the heat collection will be by one party (the owner of the tunnel) and the heat use may be by another party (e.g. neighboring private businesses, public facilities and residents). In addition to identifying that the responsibilities of each party in terms of provision of access and system maintenance has to be clearly defined on a legal basis, [FRO 10] also suggests that there is a potential income stream for the tunnel owner, either as an energy supplier or in leasing the energy rights.

Physical access also has to be considered; there will be a trade-off in terms of the optimal circuit length (before energy losses in the pipe network become unacceptably high), where distributor pipes are able to enter and exit the tunnel system from the ground, i.e. at access shafts and station platforms, and where potential end-users are located.

Nicholson *et al.* [NIC 13] describe the assessment made at "city scale" in terms of the distribution of access points and potential demand for energy harvested by energy geostructures along the route of the Crossrail project in London (Figure 12.18). Clearly, for this type of scheme to work on a large scale, vision and planning from an early stage are required in order for supply and demand to be properly and economically matched.

Thermally activated foundation elements are located within the footprint of a development, and the potential impact of HX on neighbors is likely to be minimal (except perhaps where there are significant groundwater flows and where thermally activated wall elements are at the boundary of properties). Tunnels, however, pass below the footprint of multiple landowners.

Therefore, in addition to the physical effect of the tunneling, the harvesting of energy from the ground below properties may require special permission – this may involve a demonstration that the thermal loading will not adversely affect the ground and hence the property (as with the effect of mechanical loss of ground during tunneling) and potentially, could also entail compensation for the use of the thermal resource that will be "lost" to the landowner.

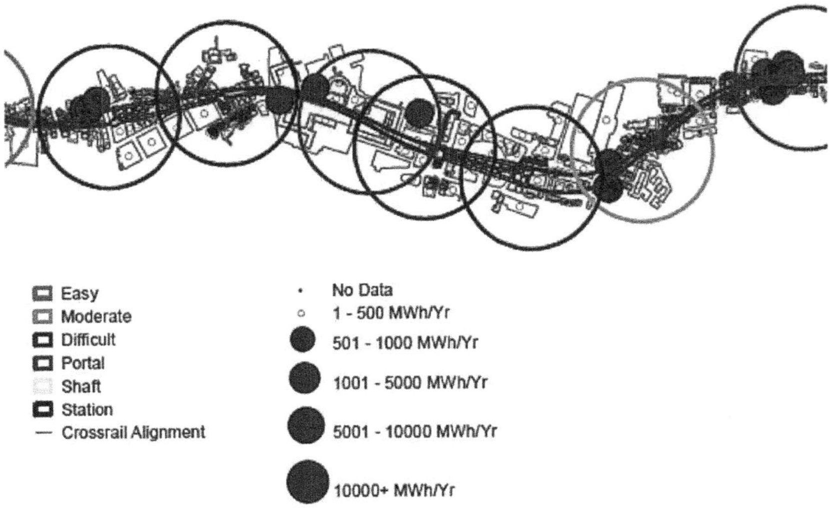

Figure 12.18. *City scale mapping of heat demand and tunnel access locations along part of London's Crossrail route, from [NIC 13]*

12.4. System integration and commissioning

The successful delivery of an energy geostructure (and in fact, BHX systems) goes beyond the requirements for the physical installation of the system. Operationally, GSHP systems are in some ways more complex than conventional heating and cooling systems, and where they are part of a wider building climate control system, (i.e. they are used in combination with conventional condensing boilers or district heating schemes), a period of time will be required in order to optimize the use of the renewable energy source to obtain best value.

To enable such tuning of the system, it will be necessary to install a monitoring system (perhaps integrated with the building management system, BMS) to record key operational variables such as the actual building heating and cooling demand profile, the energy supplied by each energy source to meet demand, temperatures within ground loops, energy supply to and flow-rates within heat pumps.

The need for this ongoing involvement during early commissioning of GSHP systems is demonstrated in the study reported by [SCH 06] and [KIP 09] – see also Chapter 3. The study reviewed the operational performance of three borehole heat exchange systems, two energy pile systems and a thermally-activated raft. Half of the systems were found to be underperforming with seasonal performance factors (SPFs) of less than three. Over the three year period of the study, in co-operation with the building owners, the SPFs were improved by one-third to two times the initial values to yield SPFs in excess of three and as high as eight, with some potential for further optimization via greater use of "free cooling".

The reasons for the underperformance identified by Schnürer *et al.* [SCH 06] and Kipry *et al.* [KIP 09] includes the following:

– The small temperature differences involved which means that a poorly designed and regulated system may not extract heat in winter / inject heat in summer sufficient to maintain operation through the opposing season.

– The reaction time is such that without adequate control systems, faster conventional thermal systems may override the GSHP system.

– Because of the lag in system response, strong seasonal fluctuations in climatic conditions in spring and autumn may lead to unnecessary system operation without appropriate system management criteria such as the definition of an external air temperature range within which neither heating nor cooling of the building is activated [KIP 09].

12.5. Summary

At the heart of the successful delivery of an energy geostructure, are the dual requirements of installing the HX elements without damage and with minimal delay (and cost) to the conventional construction process. This chapter has attempted to highlight these areas that need particular consideration in this process, from the preconstruction phase, all the way through to commissioning:

– Coordination at all stages of the delivery process is the key to success and the responsibility for undertaking this activity should be clearly defined.

– Energy provision from the GSHP system should be reviewed throughout the delivery process, to ensure all design and construction modifications are accounted for and the design energy demand is met. This is also necessary as a baseline for system optimization once operational.

– During construction, the pipe work for the GSHP system will interact with a number of different trades and the handover between these trades must be managed well. Visual inspections, pressure and flow testing should be contractual obligations at these times.

– Loss of function within the HX circuits must be kept to a minimum and contractual responsibilities for any loss need to be clearly defined.

– Many different types of pile, wall and tunnel element have now been thermally activated for the use in GSHP systems, and a wide range of details have been developed, as illustrated in the case studies presented.

– The potential for leakage where pipe work penetrates the building envelope must be addressed and appropriate waterproofing details should be used.

– Foundation and wall energy systems tend to be operated on a site-by-site basis, and in some circumstances there may be a thermal impact on neighboring properties. The management of the affected stakeholder should however be no more onerous than typical party wall approvals for conventional developments.

– Linear energy geostructures, such as tunnels, encroach on multiple properties and the planning for and delivery of such systems will require a higher level of planning and stakeholder management than a conventional building development, for example. Again this is typical of such projects in any case, although the energy planning side is a new aspect.

– Most GSHP systems will require a period of operational optimization in order to ensure that control systems maximize the performance of the renewable heating and cooling element. This will be particularly true when the GSHP system is operated alongside more conventional systems.

The case studies that appear in the chapter not only highlight the aspects of the process summarized earlier but they also represent examples of a variety of energy geostructures, which have been delivered successfully and which will contribute to the reduction in CO_2 emissions and running costs of the end user's building throughout its design life.

12.6. Acknowledgments

The lead author would like to thank his co-authors, who have all make time in their business days in order to provide their input to this chapter. Further thanks must also be extended to their companies Bilfinger, ECOME, Ed Züblin, ENERCRET, GI Energy, and Ruukki for permitting them to contribute. Many thanks also to everyone who gave their permission for the use of the images as acknowledged in the text.

12.7. Bibliography

[AMI 10] AMIS T., ROBINSON C.A.W., WONG S., "Integrating geothermal loops into the diaphragm walls of the Knightsbridge Palace Hotel project", *Proceedings of the 11th DFI/EFFC International Conference on Geotechnical Challenges in Urban Regeneration*, London, p. 10, 2010.

[BRA 06] BRANDL H., "Energy foundations and other thermo-active ground structures", *Géotechnique*, vol. 56, no. 2, pp. 81–122, 2006.

[FRA 11] FRANZIUS J.N., PRALLE N., "Turning segmental tunnels into sources of renewable energy", *Proceedings of the ICE Civil Engineering*, vol. 164, no. 1, pp. 35–40, 2011.

[FRO 10] FRODL S., FRANZIUS J.K., BARTL T., "Design and construction of the tunnel geothermal system in Jenbach", *Geomechanics and Tunnelling*, vol. 3, no. 5, pp. 658–668, 2010.

[GRO 12] GROUND SOURCE HEAT PUMP ASSOCIATION, "Thermal pile design installation and materials standards, Issue 1.0", Ground Source Heat Pump Association, Milton Keynes, UK, p. 85, available at www.gshp.org.uk/Standards.html, 2012.

[HUD 10] HUDE N., WEGNER W., "Energy piles as interface between foundation engineering and building services", *BauPortal*, vol. 2, 2010, pp. 2–7, [original in German], available at www.building-construction-machinery.net/shop/topics/special-civil-engineering/topic/. [English version], 2010.

[KIP 09] KIPRY H., BOCKELMANN F., PLESSER S., *et al.*, "Evaluation and optimization of UTES systems of energy efficient office buildings (WKSP)", *Proceedings of the 11th International Conference on Thermal Energy Storage*, Paper 43, EFFSTOCK, Stockholm, available at http://intraweb.stockton.edu/eyos/energy_studies/content/docs/effstock09/Session_6_1_Case_studies_residential_and_commercia_buildings/43.pdf, 2009.

[NIC 13] NICHOLSON D.P., CHEN Q., PILLAI A., *et al.*, "Developments in thermal pile and thermal tunnel linings for city scale GSHP systems", *Proceedings of the 38th Workshop on Geothermal Reservoir Engineering*, Stanford University, CA, Stanford, p. 8, available at https://pangea.stanford.edu/ERE/pdf/IGAstandard/SGW/2013/Nicholson.pdf, 2013.

[PRA 09] PRALLE N., FRANZIUS J.-N., ACOSTA F., *et al.*, "Using tunneling concrete segments as geothermal energy collectors", *Proceedings of the Central European Congress on Concrete Engineering*, Baden, pp. 137–141, 2009.

[SCH 06] SCHNÜRER H., SASSE C., FISCH M.N., "Thermal energy storage in office |buildings foundations", *Proceedings of the 10th International Conference on Thermal Energy Storage*, ECOSTOCK, Galloway, NJ, available at http://intraweb.stockton.edu/eyos/energy_studies/content/docs/ FINAL_PAPERS/11A-4.pdf, 2006.

[SCH 10] SCHNEIDER M., MOORMANN C., "GeoTU6 – a geothermal Research Project for Tunnels", *Tunnel*, vol. 29, no. 2, pp. 14–21, available at www.uni-stuttgart.de/igs/content/publications/190.pdf, 2010.

[UNT 04] UNTERBERGER W., HOFINGER H., GRÜNSTÄUDL T., *et al.*, "Utilization of tunnels as source of ground heat and cooling – practical applications in Austria", *Proceedings of the 3rd ISRM International Symposium*, ARMS, Kyoto, pp. 421–426, 2004.

[UOT 12] UOTINEN V.-M., REPO T., VESAMAKI H., "Energy piles – ground energy system integrated to the steel foundation piles", *Proceedings of the 16th Nordic Geotechnical Meeting (NGM 2012)*, Copenhagen, pp. 837–844, 2012.

[WIL 03] WILHELM J., RYBACK L., "The geothermal potential of Swiss Alpine tunnels", *Geothermics*, vol. 32, nos. 4–6, pp. 557–568, 2003.

Chapter 13

Thermo-Pile: A Numerical Tool for the Design of Energy Piles

13.1. Basic assumptions

Thermo-Pile was developed to estimate the effects of temperature variations on stresses and strains within pile foundations. Thermo-Pile considers one single axially loaded pile at a time whose cross-section is circular and which is embedded within a layered soil. Modeled piles are first subjected to a mechanical load and then to a monotonic temperature variation (i.e. increase or decrease), the cyclic aspects of the design being developed at the moment of writing this chapter. The temperature variation can be specified as a single value – constant along the pile axis – or as a temperature profile based on *in situ* measurements.

The thermomechanical response of the pile is assumed to be thermoelastic and is time-independent. The soil response is modeled with the load-transfer method. It uses elastoplastic load-transfer curves linking the pile displacements to the mobilized bearing capacities. However, load-transfer curves are assumed not to be influenced by the temperature variations so that no heat transport is modeled within the soil. Negative friction coming from settlements of the soil around the pile is not taken into account and neither are thermal radial strains of the pile. The reaction of the supported structure (raft, wall, etc.) is modeled with a linear elastic stiffness that reacts to the pile head heave or settlement after mechanical loading. As a result, the mechanical loading is an initialization to the thermal effects.

Chapter written by Thomas MIMOUNI and Lyesse LALOUI.

13.2. Mathematical formulation and numerical implementation

13.2.1. *The load-transfer method*

The load-transfer method uses load-transfer curves to represent the soil response. These curves link the displacements of the pile to the mobilized bearing capacities (i.e. base compression and shaft resistance).

13.2.1.1. *Shape of the load-transfer curves*

Different load-transfer functions are available in the literature [ARM 87, FRA 82, FRA 91, RAN 78] but Thermo-Pile uses the curves proposed by Frank and Zhao [FRA 82], where the slopes K_s and K_b of the load-transfer curves for the shaft friction and base compression, respectively, are estimated from the Menard pressuremeter coefficient E_M. For cohesive soils, the method proposes:

$$K_S = \frac{2E_M}{D} \text{ and } K_b = \frac{11E_M}{D} \qquad [13.1]$$

where D is the pile diameter. The same parameters for granular soils are:

$$K_S = \frac{0.8E_M}{D} \text{ and } K_b = \frac{4.8E_M}{D} \qquad [13.2]$$

The Frank and Zhao curves are made up of two slopes. The elastic branch the slope of which is K_s (K_b) extends from 0 to $\pm q_s/2$ ($\pm q_b/2$). Next, the curve follows a trend equal to the fifth of the elastic modulus until it reaches a plateau value. This particular model is shown in Figure 13.1(b). In all models, unloading is achieved according to an elastic branch.

Other models can be defined manually (Figures 13.1(a) and (c)).

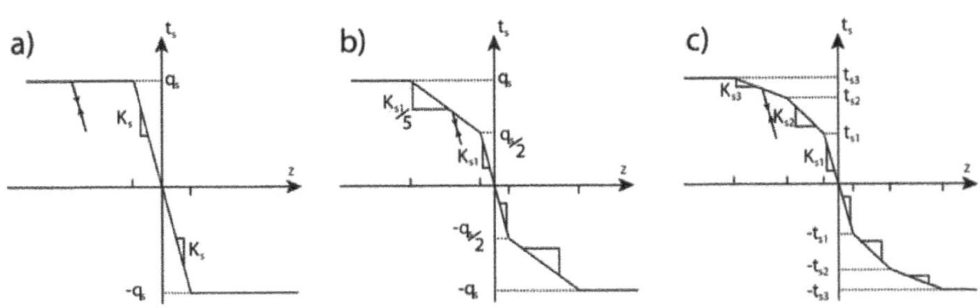

Figure 13.1. *Examples of load-transfer curves for shaft friction implemented in Thermo-Pile*

13.2.1.2. *Ultimate bearing capacities*

Thermo-Pile proposes four methods to estimate the ultimate bearing capacities of soils (i.e. the plateau values). The first two methods are based on the DTU (*Document Technique Unifié*, French rules) for granular soils: DTU empiric and DTU analytic. The third method is based on the theory of Lang and Huder [LAN 78]. Finally, the user can specify manually the ultimate bearing capacities q_s and q_b. Parameters used in these methods are listed in Table 13.1.

Symbol	Parameter
c'	Cohesion of the soil
δ	Interface friction angle
φ'	Internal friction angle of the soil
k	Lateral earth pressure coefficient

Table 13.1. *List of the different physical parameters used in the different methods*

DTU method: for the DTU method, the base resistance is estimated with the same method for empirical and analytical formulations from:

$$\begin{cases} q_b = 50N_q + \lambda N_c c' \\ N_q = 10^{3.04\tan\varphi'} \text{ and } N_c = \left(N_q - 1\right)\tan^{-1}\left(\varphi'\right) \end{cases} \qquad [13.3]$$

where N_c and N_q are bearing factors and λ is a factor of shape of the pile, equal to 1.3 for piles with a circular cross-section. The ultimate shaft friction is user-specified for the DTU empiric and calculated for the DTU analytic method with:

$$\begin{cases} q_s = k\sigma_v' \tan\delta \\ k = k_0 = 1 - \sin\varphi' \end{cases} \qquad [13.4]$$

where σ_v' is the vertical stress profile along the pile shaft.

Lang and Huder theory: the method proposed by Lang and Huder [LAN 78] estimates the ultimate base resistance based on:

$$\begin{cases} q_b = \left(N_c c' + N_q \sigma_v'\right)\chi \\ N_q = \exp\left(\pi.\tan\varphi'\right).\tan^2\left(\dfrac{\pi}{2} + \dfrac{\varphi'}{2}\right) \text{ and } N_c = \left(N_q - 1\right)\tan^{-1}\left(\varphi'\right) \end{cases} \qquad [13.5]$$

where N_c and N_q are bearing factors, $\sigma_v^{'}$ is the vertical stress in the soil at the pile base level and χ is a correction factor depending on the shape and length of the pile.

The ultimate friction at the soil–pile interface is estimated from:

$$\begin{cases} q_s = c^{'} + \sigma_v^{'}.k.\tan\delta \\ k = k_0 = 1 - \sin\varphi^{'} \end{cases} \qquad [13.6]$$

13.2.2. *Displacements induced by the mechanical load*

Displacements of the pile under a dead-load P are estimated with the load-transfer method described by Coyle and Reese [COY 66]. Let us consider the ith element of the pile as shown in Figure 13.2, the diameter of which is D and length is h_i. External forces acting on this element are the axial forces at its head ($z_{H,i}$) and base ($z_{B,i}$), $F_{H,i}$ and $F_{B,i}$, respectively, plus the shaft friction, represented by the friction acting at the middle of it $t_{s,i}(z_{M,i})$. $F_{M,i}$ is the axial force acting at the middle of the element ($z_{M,i}$). Applying Hooke's law to the lower half of the element leads to:

$$\begin{aligned} \Delta z_i &= \frac{F_{B,i} + F_{M,i}}{2} \frac{1}{AE} \frac{h_i}{2} \\ &= \left(F_{B,i} + \frac{1}{2} \frac{h_i D\pi}{2} t_{s,i}\left(z_{M,i}\right) \right) \frac{1}{AE} \frac{h_i}{2} \end{aligned} \qquad [13.7]$$

with $\Delta z_i = z_{M,i} - z_{B,i}$, we obtain:

$$z_{M,i} - z_{B,i} - \left(F_{B,i} + \frac{1}{2} \frac{h_i D\pi}{2} t_{s,i}\left(z_{M,i}\right) \right) \frac{1}{AE} \frac{h_i}{2} = 0 \qquad [13.8]$$

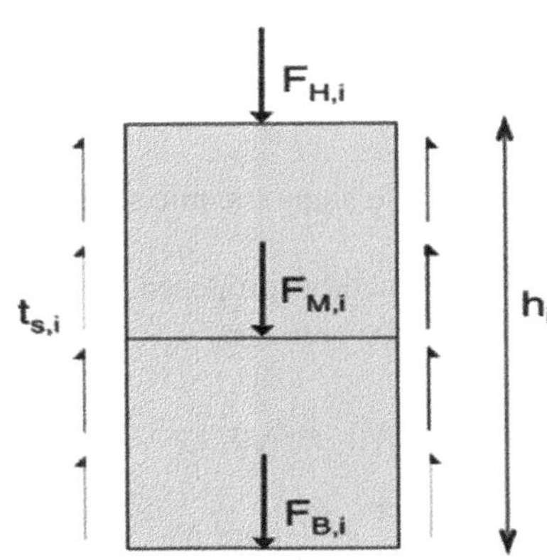

Figure 13.2. *Sketch of a pile element, from [KNE 11]*

An estimate of $z_{M,i}$ is obtained by solving iteratively equation [13.8]. Then, $F_{M;i}$ is deduced with:

$$F_{M,i} = F_{B,i} + \frac{h_i D\pi}{2} t_{s,i}\left(z_{M,i}\right) \quad [13.9]$$

Finally, the force $F_{H,i}$, at the head of the element, is given by:

$$F_{H,i} = F_{B,i} + 2\left(F_{M,i} - F_{B,i}\right) \quad [13.10]$$

and $z_{H,i}$, the corresponding displacement, is:

$$z_{H,i} = z_{B,i} + \frac{F_{M,i}}{AE} h_i \quad [13.11]$$

The displacements are solved from the pile base to the pile top because the displacement estimated at the head of an element is used as the base displacement of the overlying element. The initial displacement, Δz_n, at the base of the pile has to be chosen so that the static equilibrium of the pile is fulfilled:

$$T_b + \sum_{i=1}^{n} T_{s,i} + P = 0 \quad [13.12]$$

Axial strains induced by the mechanical load are finally deduced from axial forces within the pile.

13.2.3. *Displacements induced by the thermal load*

When a pile is heated or cooled, it expands or contracts around the null point [BOU 09] that does not move under thermal loading. As a result, the sum of the thermally mobilized forces above the null point is equal to the sum of mobilized forces below it:

$$\sum T_{th,NP} = \sum_{i=1}^{NP} T_{s,th,i} + T_{h,th} + \sum_{i=NP+1}^{n} T_{s,th,i} + T_{b,th} = 0 \quad [13.13]$$

13.2.3.1. *Case without mechanical loading*

We first consider the particular case of heating and cooling without mechanical load. In this case, there is no strain prior to temperature change. An iterative procedure is used to estimate the blocked strain.

1) *Initial strain profile*: the pile is initially assumed to be totally free to compute a first set of mobilized resistance (mobilized shaft friction and resistance at the extremities). Therefore, the initial strains are equal to the free strain: $\varepsilon_{th} = \varepsilon_{th,f} = \alpha\Delta T$.

2) *Profile of displacements*: by definition, there is no displacement at the null point ($z_{th,NP} = 0$). Therefore, displacements are cumulated from the null point toward pile ends. From these displacements, *t-z* curves provide a first set of mobilized reaction stresses along the pile axis:

$$\sigma_{th,i} = \left(T_{th,b} + h_i D\pi \sum_{j=1}^{n} t_{th,s,j} \right) \frac{1}{A} \qquad [13.14]$$

3) Blocked thermal strains $\varepsilon_{th,b}$ are therefore estimated with:

$$\varepsilon_{th,d} = \frac{\sigma_{th}}{E} \leq \varepsilon_{th,f} \qquad [13.15]$$

4) The real (i.e. observed) strain is finally estimated by subtracting the blocked strain to the free strain: $\varepsilon_{th,o} = \varepsilon_{th,f} - \varepsilon_{th,d}$.

Steps 2–4 are repeated with the new set of observed strains, $\varepsilon_{th} = \varepsilon_{th,o}$ until convergence. Then, related quantities, such as pile displacement, internal axial stresses, mobilized shaft friction and mobilized reaction at the base and head of the pile are estimated.

13.2.3.2. *Case with mechanical loading*

In this case, the mechanical loading acts as an initialization to thermal effects. In the case of unloading (uplift), the stress path follows the unloading branch.

13.3. Validation of the method

Thermo-Pile was validated against two experimental full-scale *in situ* experiments [KNE 11]: the EPFL test pile under a real building [LAL 03] and the Lambeth College test pile [BOU 09].

The EPFL test pile: modeled and measured strain profiles along the EPFL test pile during heating are shown in Figure 13.3. The agreement between the *in situ* measurements and the modeled values is good. However, a constant profile of temperature equal to the average increase in temperature within the pile was used for the simulations, instead of the measured profiles. As a result, overestimates can be observed at the pile ends where the temperature is lower than the average.

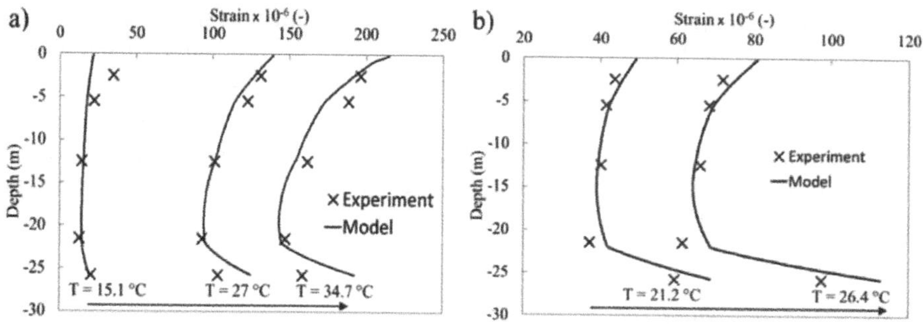

Figure 13.3. *Modeled and measured strains while heating during a) test 1 and b) test 7 on the EPFL test pile*

Soil layer	A1	A2	B	C	D
q_s (kPa)	102	70	74	160	–
q_b (MPa)	–	–	–	–	11
K_s (kPa/m)	16.7	10.8	18.2	121.4	–
K_b (kPa/m)	–	–	–	–	667.7

Table 13.2. *Soil parameters used for modeling the EPFL test pile*

Soil layer	1	2	3	4
Depth (m)	0–6.5	6.5–10.5	10.5–16.5	16.5–22.5
E_M (MPa)	45	45	45	45
q_s (kPa)	35	60	70	80
q_b (kPa)	–	–	–	460

Table 13.3. *Soil parameters used for modeling the Lambeth College test pile*

13.4. Piled-beams with energy piles

Interactions between energy piles through the rigidity of the supported structure can lead to significant variations in the behavior of the foundation. Indeed, heating or cooling a whole piled foundation induces group movements that reduce the pile–structure–pile interactions by minimizing the differential heaves or settlements. Conversely, heating only a part of a piled foundation leads to greater differential movements between the heated piles and the isothermal ones, inducing greater interactions [DUP 13].

In situ measurements have shown that a major component of these interactions is the raft. Laloui *et al.* [LAL 99, LAL 03] have shown that the degree of freedom of the test piles at EPFL decreased from 0.8 to 0.6 after the construction of the base floor, whereas it only decreased from 0.6 to 0.5 after the construction of the four floors. As a result, understanding the impact of rafts and beams on the behavior of the piles is an important component of their design because differential settlements should be minimized.

13.4.1. *General method*

The present section presents a novel method to investigate elementary piled beams with energy piles. This method couples a design tool for energy piles (Thermo-Pile) with the Euler-Bernoulli beam model [BAU 09]. The energy piles are represented as punctual forces acting on the beam, and the isothermal piles are represented by inelastic supports. The reaction of the beam to the pile head movements is modelled with head stiffness K_h in Thermo-Pile [KNE 11]. The bearing of the soil on the raft is neglected at first and remains conservative. The method is only described for isostatic cases that can be simply represented by two static piles at the beam ends and with one energy pile in between (Figure 13.4).

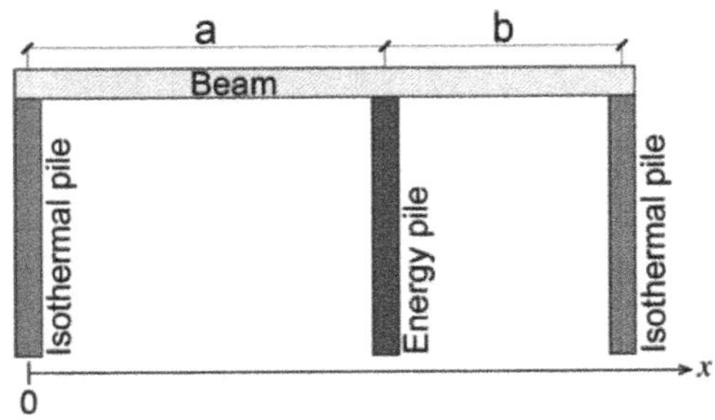

Figure 13.4. *Schematic of the pile beam with two isothermal piles at the ends and one energy pile in between*

The energy pile (Figure 13.4) expands or contracts under temperature variations. As a result, its head heaves or settles, inducing a reaction of the beam that is transmitted to the isothermal piles. Let R_P, R_L and R_R be the actions applied to the beam at the energy pile, left pile and right pile, respectively. At first, we consider that no mechanical load is initially applied to the beam so that before any thermal load, $P_P = P_L = P_R = 0$. The beam distributes the action R_P due to the expansion of the energy pile to the left and right piles according to its static equilibrium (Figure 13.5), given by:

$$\left.\begin{aligned} R_P - R_L - R_R &= 0 \\ a.R_L - b.R_R &= 0 \end{aligned}\right\} \Rightarrow \begin{cases} R_L = \dfrac{a}{L} R_P \\ R_R = \dfrac{b}{L} R_P \end{cases} \quad [13.16]$$

where a and b define the position of the energy pile below the beam (Figure 13.4) and L (=a+b) is the beam length. Therefore, an iterative process is required to solve the problem. The present approach is based on the following sequence:

1) Initialization is achieved by assuming the two isothermal piles to be fixed supports; the pile head heaves v_L and v_R of the left and right piles, respectively, are set to 0. The static of the pile then leads to a relationship linking the beam deflection v_P at the point of application of the force to the force magnitude:

$$R_P = \frac{3EI_{Gz}L}{a^2b^2} v_P = -P_P = K_h v_P \quad [13.17]$$

where I_{Gz} is the quadratic moment of the beam. As a result, the static of the beam leads to an initial head stiffness value given by:

$$K_h = \frac{3EI_{Gz}L}{a^2b^2} \quad [13.18]$$

2) The first analysis is conducted with Thermo-Pile using the value of K_h found in equation [13.18]. This analysis provides a head displacement v_P of the energy pile. From this, a first action R_P applied to the pile is estimated with [13.17]. This force is then distributed to the left and right piles with equation [13.16].

3) Two analyses are conducted with Thermo-Pile under isothermal conditions for the left and right piles with the actions previously determined. Those analyses provide two new head displacements of the isothermal piles, v_L and v_R.

4) The head displacements at the isothermal piles are used to estimate a new value of K_h. This step is achieved by determining the new set of integration constants used to estimate the deflection profile of the beam (see equation [12.25]). The new effective displacement v_P is then estimated with the new constants. From this, a new value for the head stiffness is found by dividing the action R_P by the new value of v_P.

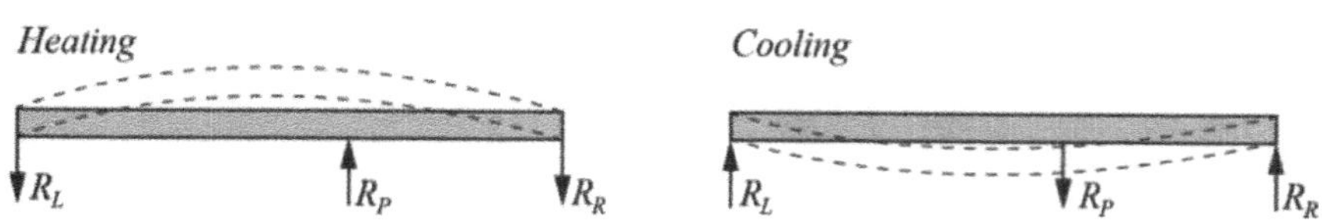

Figure 13.5. *Representation of the forces applied to the beam by the piles while heating or cooling the energy pile*

Steps 2 to 4 are then repeated until convergence is reached. A schematic of the algorithm is given in Figure 13.6.

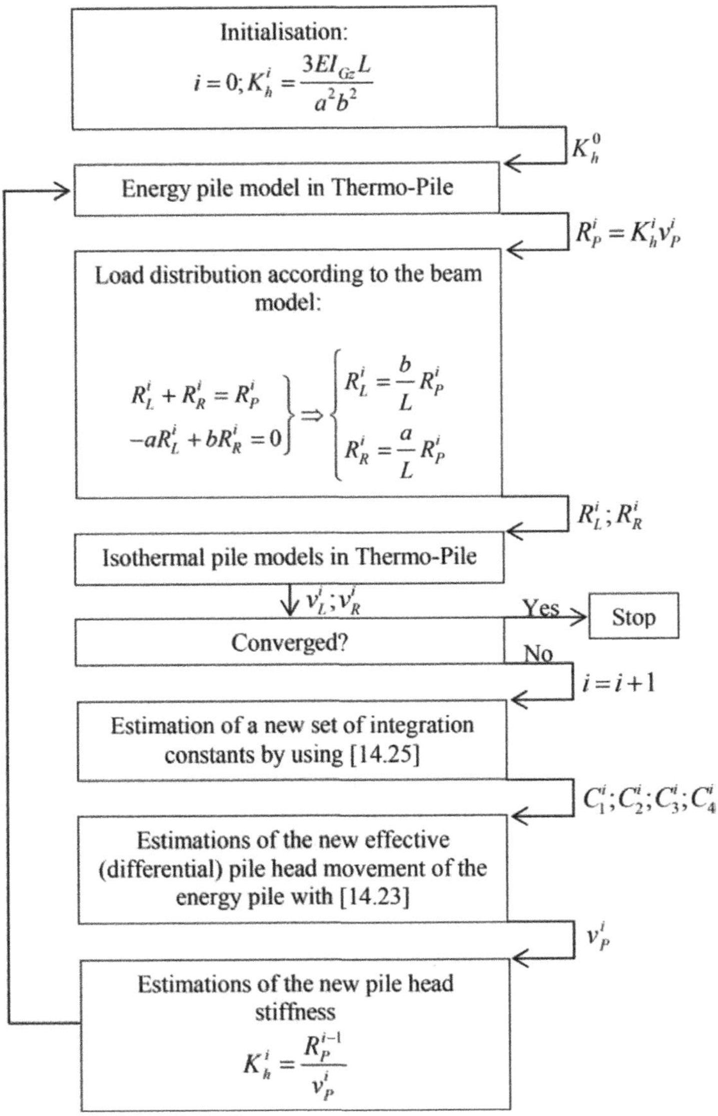

Figure 13.6. *Detailed iterative algorithm*

13.4.2. *Determination of the integration constants*

Step 4 is detailed in the following paragraph. The method employed is based on Euler-Bernoulli beam theory, and we consider a simply supported beam with a punctual load, which is given here as an example (Figure 13.6).

Static equilibrium of the beam yields:

$$R_A = \frac{b}{L} P \text{ and } R_B = \frac{a}{L} P \qquad [13.19]$$

where R_A and R_B are the support reactions at points A and B, respectively. The relationship between the curvature of the beam and the bending moment M_{fz} assuming small deflections leads to:

$$EI_{Gz} v'' = M_{fz}(X) \qquad [13.20]$$

where v is the beam deflection and X the longitudinal coordination along the beam from A ($X = 0$) to B ($X = L$). For a cross-section between A and C (Figure 13.7), the bending moment is:

$$M_{fz1}(X) = R_A X = \frac{b}{L} PX \qquad [13.21]$$

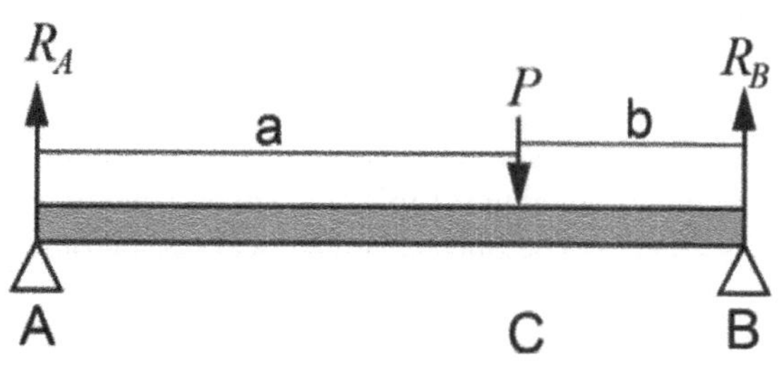

Figure 13.7. *Schematic of a generic simply supported isostatic beam*

Using equations [13.20] and [13.21] with equation [13.19] and integrating two times yields:

$$EI_{Gz} v_1(X) = \frac{bP}{6L} X^3 + C_1 X + C_2$$
$$EI_{Gz} v_2(X) = \frac{aP}{6L} (L - X)^3 + C_3 X + C_4 \qquad [13.22]$$

where C_1, C_2, C_3 and C_4 are the constants of integration. Finally, the boundary conditions of the problem are:

$$\begin{aligned} & v_1(X=0)=v_L \\ & v_2(X=L)=v_R \\ & v_1(X=a)=v_2(X=a) \\ & v_1'(X=a)=v_2'(X=a) \end{aligned} \qquad [13.23]$$

Inserting these boundary conditions in equation [13.22] leads to the system:

$$\begin{bmatrix} 0 & 1 & 0 & 0 \\ 0 & 0 & L & 1 \\ a & 1 & -a & -1 \\ 1 & 0 & -1 & 0 \end{bmatrix} \begin{bmatrix} C_1 \\ C_2 \\ C_3 \\ C_4 \end{bmatrix} = \begin{bmatrix} EI_{Gz} v_L \\ EI_{Gz} v_R \\ \dfrac{ab\left(b^2-a^2\right)}{6L} P \\ -\dfrac{ab}{2} P \end{bmatrix} \qquad [13.24]$$

where the unknown vector is formed by the four constants of integration.

13.4.3. *Example of simulation*

A simple example is investigated in this section. The 5-m-long beam used in the study is assumed to have a cross-section with sides such that the quadratic moment of the beam cross section is equal to 0.1 m^4. The Young's modulus of the concrete in the beam is taken as 20 GPa. The piles have identical geometries with a diameter of 0.5 m and a length of 10 m. They are embedded in the same homogeneous soil, which is represented by its Menard coefficient E_M, equal to 60 MPa, and its ultimate shaft friction and base compression q_s and q_b, equal to 200 kPa and 4,500 kPa, respectively.The energy pile is 1 m away from the left pile, and an extreme temperature variation of +60°C is applied. The results of the simulation are given in Figure 13.8.

The maximum deflexion of the beam is not observed right at the energy pile (Figure 13.8 a) because of the beam rigidity and fixities at both ends. The profiles of the axial stress (Figure 13.8 b) and axial displacement (Figure 13.8 c) show that the heated pile undergoes compression, while the isothermal piles experience tension, the left pile being more affected because it is closer to the energy pile. Finally, the profile of the deflexion along the beam shows some differential displacements between the three piles, up to 1 mm between the heated pile and the right pile. However, these high values of the differential settlements must be weighted by the fact that the piles are really close to each other and that the temperature increase is very extreme.

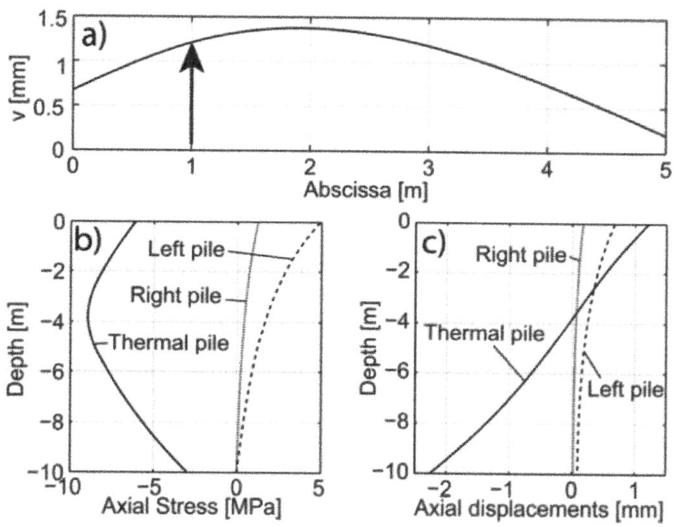

Figure 13.8. *Result of the example with a) the deformed profile of the beam, b) the profiles of the axial stress and c) the axial displacements in the piles*

13.5. Conclusions

Thermo-Pile is able to accurately reproduce measurements taken from experimental sites. The load-transfer method is a good and efficient operational method to estimate the thermal effects on energy piles under monotonic temperature variations.

The simple algorithm employed in Thermo-Pile offers great opportunities to develop more complex tools by coupling it with other models such as the Euler-Bernoulli model for beams. An example of coupling was given to study elementary piled beams with energy piles. This novel method, based on an iterative process, is easy to implement and allows real operational design compared to complex finite element methods.

13.6. Acknowledgments

The authors thank EOS Holding for their financial support through the project GRETEL II: Geotechnical Reliability of Thermo-piles Energy Storage in Soils.

13.7. Bibliography

[AMA 12] AMATYA B.L., SOGA K., BOURNE-WEBB P.J., *et al.*, "Thermo-mechanical behaviour of energy piles", *Géotechnique*, vol. 62, no. 6, pp. 503–519, 2012.

[ARM 87] ARMALEH S., DESAI C., "Load deformation response of axially loaded piles", *Journal of Geotechnical Engineering*, vol. 113, no. 12, pp. 1483–1500, 1987.

[BAU 09] BAUCHAU O.A., CRAIG J.I., "Euler-Bernoulli beam theory", *Structural Analysis*, Springer Neherlands, 2009.

[BOU 09] BOURNE-WEBB P. J., AMATYA B., SOGA K., *et al.*, "Energy pile test at Lambeth College, London: geotechnical and thermodynamic aspects of pile response to heat cycles", *Géotechnique*, vol. 59, no. 3, pp. 237–248, 2009.

[COY 66] COYLE H. M., REESE L. C., "Load transfer for axially loaded piles in clay", *Journal of the Soil Mechanics and Foundations Division*, vol. 92, no. 2, pp. 1–26, 1966.

[DUP 13] DUPRAY F., LALOUI L., KAZANGBA A., "Numerical analysis of seasonal heat storage in an energy pile foundation", *Computers and Geotechnics*, 2013.

[FRA 82] FRANK R., ZHAO S. R., "Estimation par les paramètres pressiométriques de l'enfoncement sous charge axiale de pieux forés dans des sols fins", *Bull Liaison Lab Ponts Chaussées*, vol. 119, pp. 17–24, 1982.

[FRA 91] FRANK R., KALTEZIOTIS N., BUSTAMANTE M., *et al.*, "Evaluation of performance of two piles using pressuremeter method", *Journal of Geotechnical Engineering*, vol. 117, no. 5, pp. 695–713, 1991.

[KNE 11] KNELLWOLF C., PERON H., LALOUI L., "Geotechnical analysis of heat exchanger piles", *Journal of Geotechnical and Geoenvironmental Engineering*, vol 137, no. 10, pp. 890–902, 2011.

[LAL 99] LALOUI L., MORENI M., STEINMANN G., *et al.*, Test en conditions réelles du comportement statique d'un pieu soumis à des sollicitations thermo-mécaniques, Swiss Federal Office of Energy (OFEN), report, 1999.

[LAL 03] LALOUI L., MORENI M., VULLIET L., "Comportement d'un pieu bi-fonction, fondation et échangeur de chaleur", *Canadian Geotechnical Journal*, vol. 40, no. 2, pp. 388–402, 2003.

[LAN 78] LANG H.J., HUDER J., *Bodenmechanik und Grundbau: Das Verhalten von Böden und die wichtigsten grundbaulichen Konzepte*, Springer, Berlin, Heidelberg, 1978.

[MIM 13] MIMOUNI T., LALOUI L., "Towards a secure basis for the design of geothermal piles", *Acta Geotechnica*, 2013.

[RAN 78] RANDOLPH M.F., WROTH C.P., "Analysis of deformation of vertically loaded piles", *Journal of the Geotechnical Engineering Division*, vol. 104, no. GT12, pp. 1465–1488, 1978.

Chapter 14

A Case Study: The Dock Midfield of Zurich Airport

14.1. The Dock Midfield

Dock Midfield is the new Terminal E at Zürich airport. Designed for 26 planes, the building is 500 m long and 30 m wide. It is built on 440 foundation piles because the upper ground layer, which is composed of lake deposits, is too soft to support the loads of the building without major settlements. The piles stand on moraine, which lies at a depth of about 30 m. With a diameter of 0.9–1.5 m, the concrete piles were cast in place. An image of the building is shown in Figure 14.1.

Renewable energies are used extensively throughout this building. They are expected to meet 65% and 70% of the heating and cooling requirements, respectively. The foundation piles are used as energy piles: approximately 300 piles have been equipped with five U-pipes fixed on the metallic reinforcement. The circulation of a heat carrier fluid through the U-pipes enables them to be used as a heat exchanger with the ground. The additional amount of energy purchased for heating is very small. The associated heating energy index, defined by the ratio of the annual purchased energy (district heating energy and electricity for the heat pump (HP)) per the total heated floor area (85,200 m^2 with height correction), is approximately 15 kWh/m^2y. The total electric energy index, estimated to be 110 kWh/m^2y, is also low for a fully air-conditioned building that is used 18 h a day.

Chapter written by Daniel PAHUD.

Figure 14.1. *The Dock Midfield of Zürich airport has been built on 440 foundation piles of 30 m*

The design phase of Dock Midfield was initiated in 1995. Its construction started in 2000, and the building commissioning was completed in 2003. In September 2004, the measurement of the pile system started for a two-year period.

14.2. Design process of the energy pile system

Because of its multidisciplinary character, an energy pile system concept has to be elaborated on very early in the project. This is particularly important as the foundation piles are one of the first works to be initiated. A correct sizing is required to carefully define the local geological, geotechnical and hydrogeological contexts. On the other hand, the thermal requirements of the building, both for heating and cooling, should be known in advance and with the best possible accuracy. From this point, different system variants can be assessed and compared, on the basis of the available foundation piles, their location, and their geometrical and thermal characteristics.

This procedure can be repeated several times in a project, in relation to the degree of knowledge of data and their evolution between the initial and final phases of the project. This is illustrated in the case of Dock Midfield.

14.2.1. *Pile system concept*

The energy piles allow us to combine the heating with the cooling production. Integrated in a bivalent system, they provide base heating in winter. An HP extracts heat from the piles at a relatively low thermal power, thus increasing operation duration and maximizing the extracted geothermal energy. District heating, available onsite, covers peak power loads. Cooling of the ground results from building heating. A thermal recharge is thus required for the long-term operation of the system and must be guaranteed by the cooling requirements of the building. They are satisfied by geocooling, i.e. by a direct coupling of the pile flow circuit with the cooling distribution circuit through a flat plate heat exchanger. Cooling peak power loads are covered by a cooling machine. In order not to destroy the geocooling production, the cooling machine is not coupled with the piles. Its waste heat is evacuated in cooling towers placed on the building's roof.

14.2.2. *Problems to solve*

The pile system has to be sized so that the heat carrier fluid circulating in the pipes imbedded in the piles remains in a given temperature range. This condition has to be fulfilled at any time, and particularly in the long term, after years of system operation. The temperature variations have to be compatible with the first and main function of the piles, which is to support the building weight and to provide good static properties. It is imperative that pile concrete always remains at a temperature greater than 0°C. In the case of Dock Midfield, the fluid temperature range lies between 0°C and 18°C, as the maximum temperature is conditioned by the maximum fluid temperature in the cooling distribution. With this relatively small variation to the fluid temperature, it was assumed that the effect on the static properties of the piles would be rather negligible.

The first calculations were supposed to answer the following questions:

– How many U-pipes are required per pile and what is the best position for the pipes?

– Is it better to use plain water or water and antifreeze for the heat carrier fluid?

– How much energy can be extracted and injected into the piles per year?

– What is the maximum power that can be extracted and injected into the piles?

The last two questions are related to both the thermal potential offered by the piles and the technical feasibility of the pile solution, taking into account the thermal requirements of the building. They enable the sizing of the HP. The answers are obtained by calculations and system simulations. The simulation model has to take into account:

– the local geological and hydrogeological conditions;

– the number, position, geometry and design of the energy piles;

– the heat transfer between the building base and the ground, taking into account the horizontal connecting pipes of the pile flow circuit;

– the building thermal requirements for heating and cooling;

– the temperature levels in the heating and cooling distribution systems;

– the system concept used to meet the energy requirements with the best possible integration of the energy piles for maximum thermal efficiency.

The successive calculations have refined the previous results, taking into account the lastest available information on the project. Finally, a detailed system simulation enabled us to optimize the system layout and helped to establish a control strategy of the pile system. It is interesting to retrace the history of the Dock Midfield system design. The successive steps are discussed in the following sections.

14.2.3. *First calculations*

The first calculations, performed in August 1997, were conducted in two steps. First, heat transfer calculations were made to determine the optimal number and position of U-pipes inside a pile. Then, all the piles were simulated as part of the overall system.

It was shown that U-pipes have to be placed as close as possible to the pile border and distanced as evenly as possible on the pile circumference. For practical reasons, U-pipes are fixed on the inner side of the metallic reinforcement of the concrete piles. Even for the thick piles of Dock Midfield, which have a diameter of 1.4 m, it is not necessary to install more than four vertical U-pipes.

Initial studies of hydrogeology indicated that the ground water table is close to the surface. However, no ground water movement is expected as its hydraulic gradient is practically null. Heat transfer in the ground is dominated by heat conduction, and the most influential parameter is the ground thermal conductivity, estimated to be 2 W/mK according to literature.

The first simulations were based on simple assumptions to take into account all the thermal processes involved. The first estimation of the building requirements is shown in Figure 14.2.

The first simulations were based on 196 piles of 25 m, half of them having a diameter of 1.4 m and the other half having a diameter of 1.0 m. It was shown that the pile potential was better used with water and antifreeze rather than plain water, as the minimum fluid temperature in the pile flow circuit can be decreased from 4°C to 0°C (see Table 14.1).

Because of the much higher thermal performances using antifreeze, it was decided to use water and antifreeze for the heat carrier fluid circulating in the pile flow circuit.

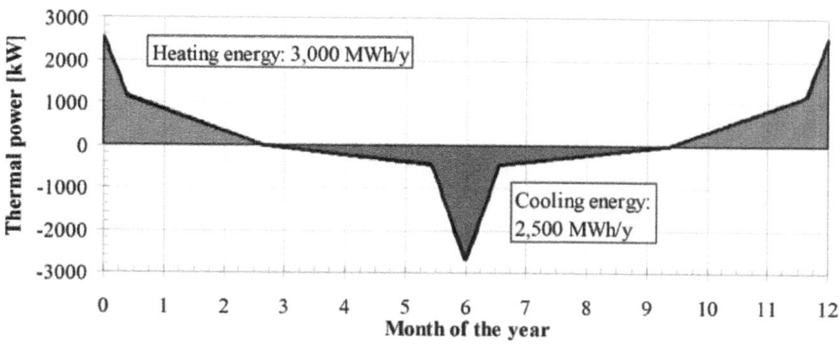

Figure 14.2. *First estimation of the heating and cooling requirements*

		Plain water	Water + antifreeze
Heating	maximum HP power annual energy	200 kW 710 MWh/y	360 kW 1,200 MWh/y
Cooling	maximum power annual energy	110 kW 490 MWh/y	150 kW 640 MWh/y
Energy piles	number – total length minimum fluid temperature extraction (winter) injection (summer)	196–4,900 m 4°C 31 W/m 108 kWh/m y 22 W/m 100 kWh/m y	196–4,900 m 0°C 55 W/m 184 kWh/m y 31 W/m 131 kWh/m y

Table 14.1. *Results of the first simulations of the pile system*

14.2.4. *Second calculations*

A second simulation was performed a couple of months later, as the thermal requirements of the building were assessed in a much more accurate way using a dynamic building simulation. The heating requirements are characterized by a peak power load of 2,800 kW and an annual energy of 1,280 MWh. The annual cooling energy is simulated to 1,200 MWh. The pile system has to cover part of it (620 MWh) with a maximum peak cooling power of 500 kW. Hourly profiles of the heating and cooling requirements were defined and used for the pile simulation (see Figure 14.3).

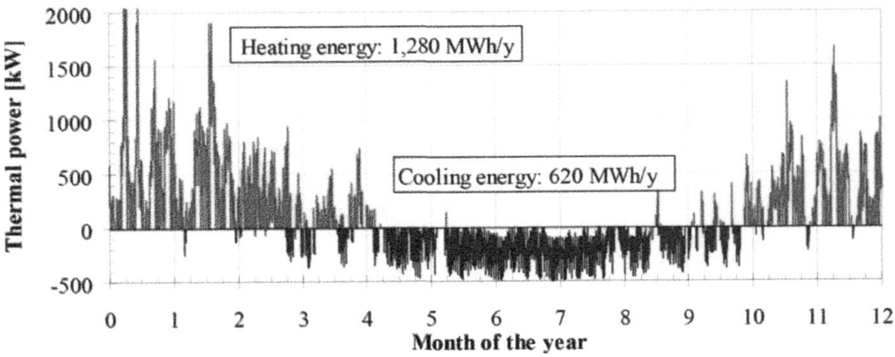

Figure 14.3. *Second assessment of the heating and cooling requirements*

The position, number and size of the foundation piles are also known clearly. At this stage of the project, 350 piles are expected, whose diameter ranges from 0.9 to 1.5 m. Their position under the building depends on the static requirements. As a result, their arrangement presents zones with a higher density of piles. From a thermal point of view, a uniform placement of the energy piles is an important criterion for the selection of the piles. It results in 304 piles that can be converted into energy piles.

The second simulations were made for 200 and 304 piles. The results are shown in Table 14.2. Because of more realistic loading conditions relative to the first calculations, it can be observed that a greater specific heat transfer rate through the piles has to be compensated for by a smaller transferred annual energy.

		200 piles variant	304 piles variant
Heating	maximum HP power annual energy	600 kW 1,030 MWh/y	800 kW 1,150 MWh/y
Cooling	maximum power annual energy	210–450 kW 580 MWh/y	360–470 kW 610 MWh/y
Energy piles	total length minimum fluid temperature extraction (winter) injection (summer)	5,000 m 0°C 86 W/m 148 kWh/m y 42–90 W/m 116 kWh/m y	7,600 m 0°C 75 W/m 108 kWh/m y 47–62 W/m 80 kWh/m y

Table 14.2. *Results of the second simulations of the pile system*

Even with 304 energy piles, the peak cooling load is not met. An auxiliary cooling power of 100–150 kW is necessary. The pile system concept is designed for this purpose: the HP can be used as a cooling machine in summer. Waste heat is not evacuated in the piles but through cooling towers on the roof.

14.2.5. *Third calculations*

The third simulation series of the system was realized one year after the second simulation series in the final design phase of the project. In the meantime, a geothermal response test was achieved to assess the local ground temperature and thermal conductivity [PAH 98a]. The heating and cooling requirements were simulated, again taking into account the last available information on the project [KOS 98]. The heating requirements were calculated to be 2,720 MWh/y with a peak power load of 4,000 kW. Part of the cooling requirements were regrouped to be coupled with the pile system. They now correspond to 1,240 MWh/y. The maximum cooling power was not affected and remained 500 kW (see Figure 14.4).

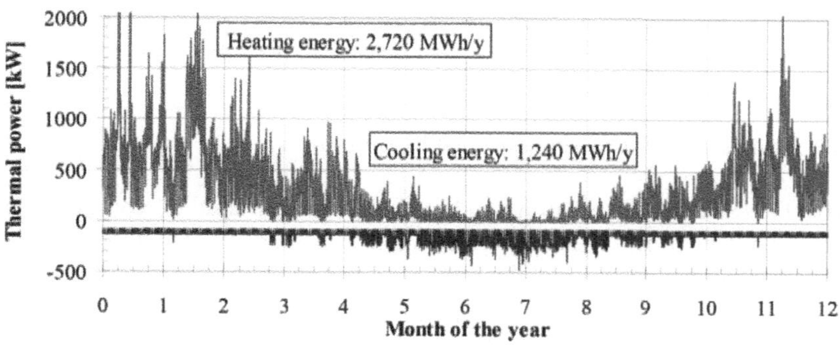

Figure 14.4. *Third and final assessment of the heating and cooling requirements*

The number, diameter, length and position of the energy piles are definitively known. The system is simulated and sized using the first version of PILESIM [PAH 98b]. The results of the simulations are shown in Table 14.3.

The heat balance of the system shows that 55% of the cooling requirements are contemporaneous to heating. They are covered by the HP for heating purposes. Geocooling energy – the cooling energy injected in the piles – represents 32% of the annual cooling requirements. The remaining 13% is covered by the HP and used as a cooling machine.

Only 36% of the annual energy extracted from the pile is injected back into the ground by geocooling. This low ground recharge ratio is only possible because of the long shape of the building (sections of 30 × 500 m) and the large spacing between the piles (approximately 9 m).

		PILESIM simulation
Heating	maximum HP power annual energy	630 kW 2,300 MWh/y
Cooling	maximum power annual energy HP piles (geocooling) cooling machine	500 kW 1,240 MWh/y 55% 32% 13%
Energy piles	number – total length minimum fluid temperature extraction (winter) injection (summer)	306–8,200 m 0°C 49 W/m 135 kWh/m y maximum 49 W/m 48 kWh/m y

Table 14.3. *Results of the third simulations of the pile system*

14.2.6. *Final simulations using the TRNSYS program*

The TRNSYS simulation program [KLE 98] has been used to reproduce the pile system layout in order to have a calculation tool that can, as close as possible, simulate the thermal behavior of the actual system. The main objectives were to:

– check HP sizing and system heat balance made using PILESIM;

– optimize the system layout;

– establish a control strategy for best system operation.

The TRNSYS simulation confirmed the HP sizing using PILESIM and provided an affined system heat balance [PAH 99].

14.3. The PILESIM program

The PILESIM program is based on the TRNSYS simulation program and TRNVDSTP [PAH 96], a non-standard TRNSYS component devised for the simulation of ground heat storage using borehole heat exchangers or energy piles. PILESIM is a TRANSED application of TRNSYS, i.e. few system layouts are possible and the user may only have access to the main design parameters through a user-friendly interface. The use of PILESIM does not require us to have a TRNSYS license or specific knowledge of TRNSYS. PILESIM was created in the framework of the Dock Midfield project to facilitate the repetitive simulations of the pile system.

PILESIM allows the user to simulate a pile or borehole system taking into account the heat transfer in the ground, in the piles or boreholes, and at the ground interface with the building. The technical installation is included in a simulation as illustrated in Figure 14.5 using the pile system border.

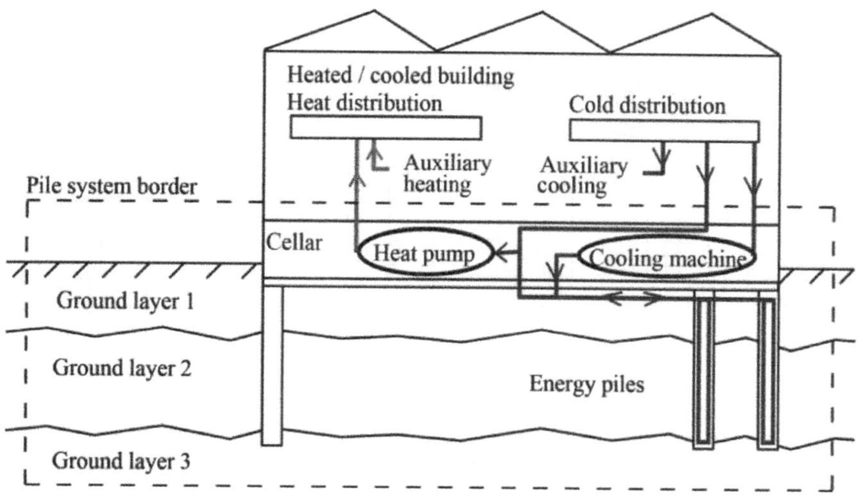

Figure 14.5. *Pile system border of the PILESIM program*

The building's heating and cooling requirements are given as input to PILESIM. Typical hourly profiles for a year have to be created or simulated using another simulation program.

PILESIM can be used in different degrees of detail, depending on the level of knowledge of the system parameters. PILESIM can be used from the early phase to the final phase of a project.

14.4. System design and measurement points

The system layout of the pile system and the measurement points are shown in Figure 14.6. Cooling energy is transferred in the pile system either through the HX-W heat exchanger or the HX-S heat exchanger as indicated in the figure. The forward fluid temperature of 14°C in the cooling distribution is controlled with a variable flow rate in the pile circuit, which is controlled with either valve V2 or V3. As the flow rate cannot be decreased below a given value, the smallest heat exchanger (HX-W) takes over the largest heat exchanger (HX-S) when the fluid temperature in the pile circuit is too low (normally in winter).

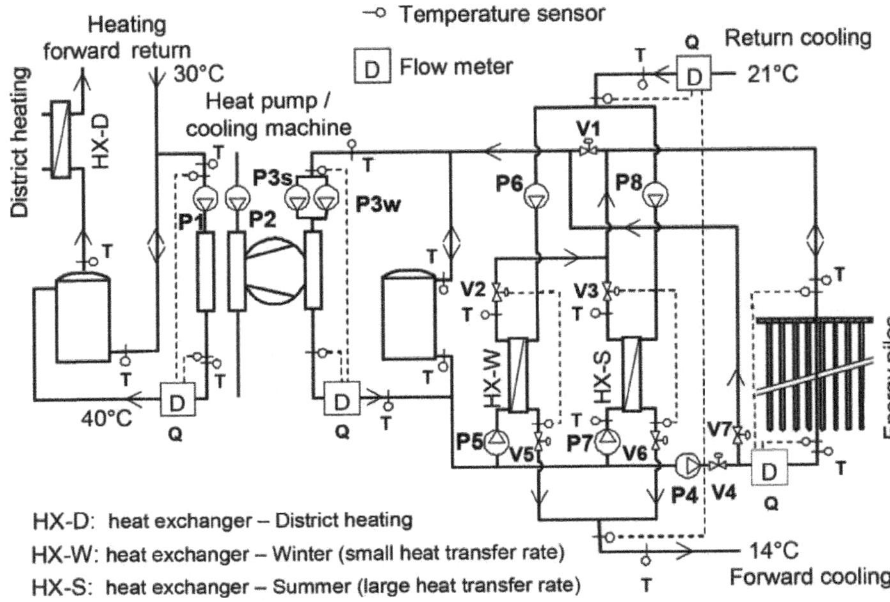

Figure 14.6. *System layout and measurement points*

The system operation mode is controlled by the on/off valves V1, V4 and V7. Heat extraction from the pile requires V1 and V4 open, V7 closed and P4 switched on. Geocooling or heat injection in the pile is achieved with V1 and V4 closed, V7 open and P4 switched off.

The pile system monitoring is performed using the measurements of 15 fluid temperatures, 11 operation status for the circulation pumps and the HP, 5 heat meters including district heating contribution, 15 ground temperatures in four piles that were not used as energy pile and the outside air temperature. These measurements are recorded by the building automation system every 5 min. Separate data loggers are also installed to record the electric consumptions of the circulation pumps, the HP and the cooling machine.

14.5. Measured thermal performances of the system

Heating energy delivered by the HP is measured to be 2,210 MWh. With a district heating contribution of 810 MWh, the annual thermal energy is measured to be 3,020 MWh. The annual thermal performance coefficient of the HP (COPA) is set as 3.9, including the electric energy for the circulation pumps P1 (condenser), P3w (evaporator) and P4 (energy piles). It would be 4.5 without the circulation pumps. The mean annual temperature level of the outlet fluid from the HP condenser is 39°C and is rather constant throughout the heating period. This suitable value is obtained because of the low return fluid temperature from the heating distribution. An average temperature of 29°C is measured, which is even better than the expected design value. Detailed results of the monitoring campaign can be found in [PAH 07a].

The measured monthly cooling energies of the cooling distribution network are shown in Figure 14.7.

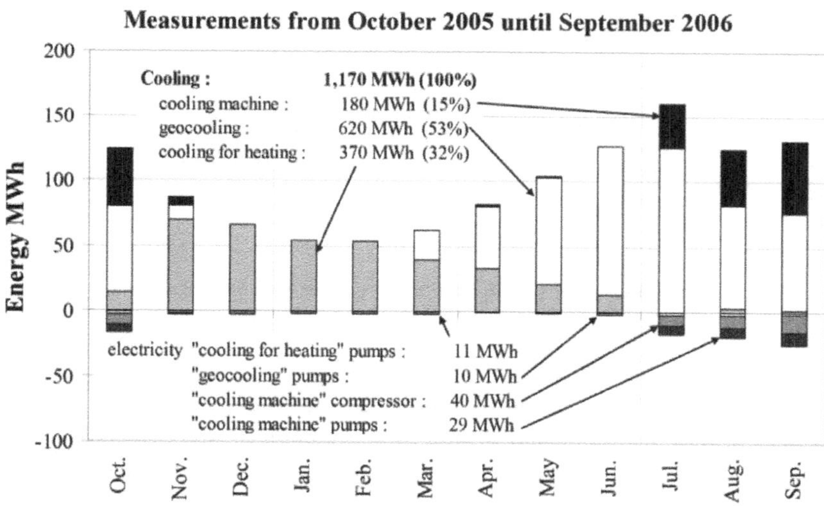

Figure 14.7. *Monthly cooling energies delivered in the cooling distribution network*

The electric energy for the circulation pumps and the cooling machine is measured to be 90 MWh. The overall cooling efficiency, defined by the ratio between the delivered cooling energy and the electric energy used to operate the system for the cooling production, is set as 13. This large value is also due to the particularly high geocooling efficiency (61). The cooling machine efficiency, set as 2.7, is hampered by the electric consumption of the circulation pumps. It represents more than 70% of the compressor electric consumption of the cooling machine. This is also an indication that the HP, used as a cooling machine, is not adequately sized relative to the cooling requirements.

The heat balance of the pile system is shown in Figure 14.8. The measured values are compared to the predicted values established using PILESIM [PAH 99].

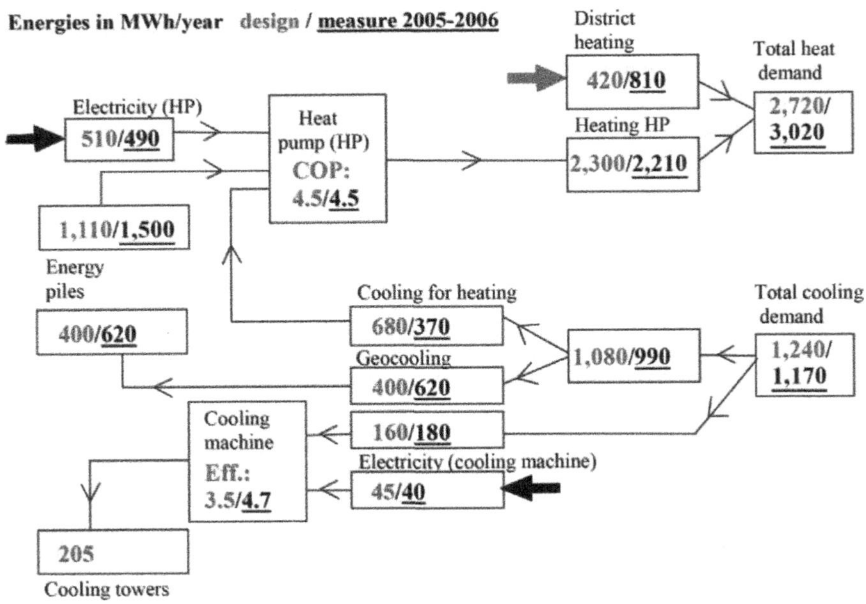

Figure 14.8. *Pile system heat balance, comparison of measured and predicted values using PILESIM*

The measured values correspond astonishingly well to the predicted values. The good accuracy of the simulated heating and cooling demands [KOS 98] can be observed. This confirms the pertinence of the design procedures adopted and the simulation tools used for the sizing of the system.

The piles are actually used more intensively than expected, as simultaneous heating and cooling requirements are not occurring as much as foreseen. However, the ground–recharge ratio is close enough to the design–recharge ratio, so that a long-term operation of the system is guaranteed.

The fluid temperature levels at the inlet and outlet of the pile flow circuit are shown with daily values for the extraction and injection operation modes (see Figure 14.9). Ground temperature is also reported. They are measured in a pile that is not used as an energy pile.

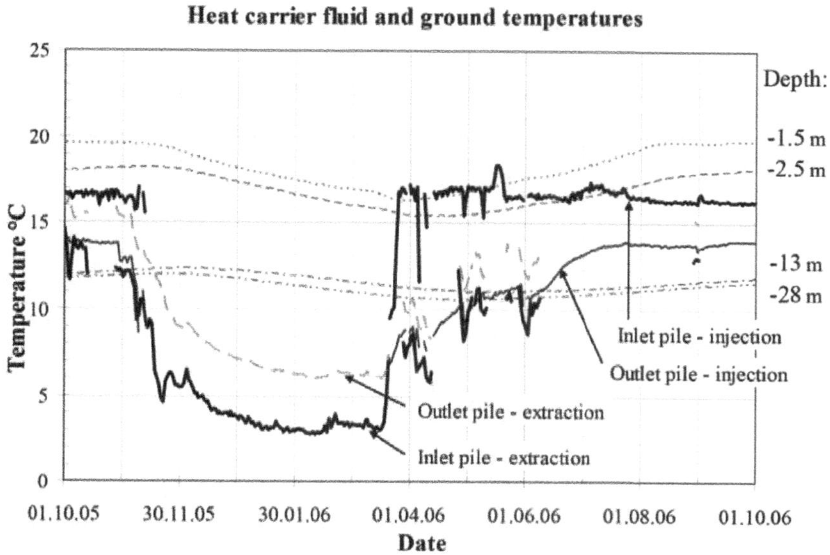

Figure 14.9. *Daily temperature levels of the heat carrier fluid in the pile flow circuit are shown for both extraction and injection operation modes. Ground temperature at various depths is measured at mid-position between energy piles*

The minimum inlet fluid temperature in the piles is measured to be 2.4 °C in the first measurement year and 2.5°C in the second measurement year. The ground temperature, below the thermal influence of the surface, exhibits seasonal but small variations due to the large ground volume involved.

The return fluid temperature from the cooling distribution, measured to be 17 °C, is lower than the expected temperature of 21 °C. The maximum heat injection rate in the pile is lower than the predicted rate and the geocooling potential is not completely used.

The overall system efficiency, defined by the ratio of the thermal energy delivered by the system (heating and cooling) and the total electric energy required to run it (all the circulation pumps, the HP and the cooling machine), is set as 5.1.

14.6. System optimization and integration

Owing to the detailed measurements of the system, the PILESIM program has been improved to better take into account geocooling calculations [PAH 07b]. It has been successfully calibrated to the measured thermal performances and was used to analyze geocooling sensitivity to various design parameters [PAH 07a].

The most important parameter is the temperature level of the fluid in the cooling distribution. In Figure 14.10, the geocooling contribution of the pile system is shown in relation to the required forward fluid temperature in the cooling distribution. The simulated cooling energy covered by geocooling and the cooling machine amounts to approximately 700 MWh/y. For various reasons, it differs slightly from the measured contribution given in Figure 14.8 (800 MWh/y).

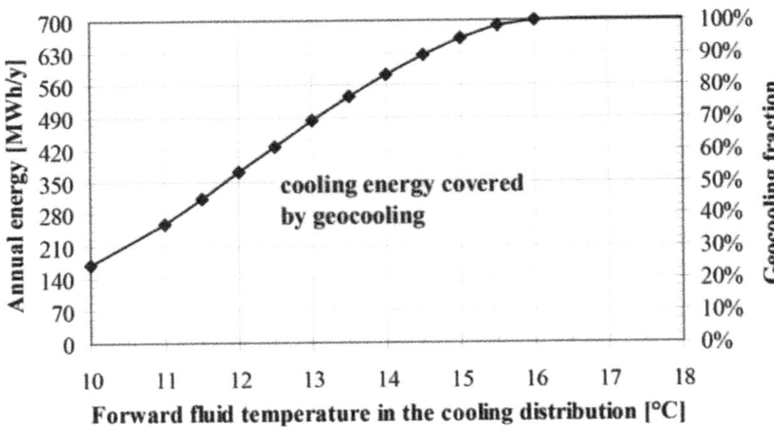

Figure 14.10. *Sensitivity of the geocooling potential to the fluid temperature level in the cooling distribution*

The geocooling potential is also strongly influenced by the horizontal pipe connections. Lying just below the building, they are exposed to a higher temperature than that of the ground (see Figure 14.9).

System integration means cooling with the highest possible temperature. This is conditioned by the size and type cooling emitters. In the case of Dock Midfield, a forward fluid temperature of 16–17°C in the cooling distribution would have prevented the cooling machine from operating. The overall system efficiency would have risen from 5.1 to 5.7. Furthermore, the probability of breakdowns would have been reduced and system reliability increased, as the system operating mode would not repeatedly switch from heating to active cooling which may occur during sunny days in October.

14.7. Conclusions

The thermal performances of the pile system are very satisfactory and they are close to the design values. It confirms the necessity of the effort put into the design phase of the system, which included, for the pile system, two thermal response tests, dynamic building simulations for the determination of the energy requirements, and pile system thermal simulations.

A very good pile system integration has been realized for the heating mode. Because of a good design and control of the heat distribution system, a low return fluid temperature has been achieved. The system concept is designed to enhance geocooling. However, the geocooling potential is extremely sensitive to the required temperature level of the distributed cooling energy.

Simulations made using PILESIM showed that the cooling machine would not be used if cooling energy is distributed at 16–17°C instead of 14°C. The overall system efficiency, measured to be 5.1, would rise to 5.7. This high efficiency is due to the geocooling efficiency, which has been measured at an exceptionally high value of 61.

Because of the good thermal performances, economical evaluations confirmed that the energy pile system of Dock Midfield is economically more profitable than a conventional pile system.

14.8. Acknowledgments

The Swiss Federal Office of Energy is acknowledged for its financial support. Unique AG and Amstein + Walthert AG are acknowledged for their indispensable collaboration and valuable contributions to the various projects involved.

14.9. Bibliography

[KLE 98] KLEIN S.A. *et al.*, TRNSYS: a transient system simulation program. Version 14.2, Solar Energy Laboratory, University of Wisconsin, Madison, WI, 1998.

[KOS 98] KOSCHENZ M., WEBER R., *Thermische Simulationsberechnungen Dock Midfield Teil 2*, EMPA, Abt. Haustechnik, Dübendorf, 1998.

[PAH 96] PAHUD D., FROMENTIN A., HADORN J.-C., The duct ground heat storage model (DST) for TRNSYS used for the simulation of heat exchanger piles, User manual, December 1996 Version, Internal Report, LASEN – DGC – EPFL, Switzerland, 1996.

[PAH 98a] PAHUD D., FROMENTIN A., HUBBUCH M., Response – Test für die Energiepfahlanlage Dock Midfield, Zürich Flughafen. Messung der Bodenleitfähigkeit in situ, Federal Office of Energy, Publication number 195323, available at www.bfe.admin.ch/dokumentation/energieforschung, 1998.

[PAH 98b] PAHUD D., PILESIM: simulation tool of heat exchanger pile systems, User manual, Laboratory of Energy Systems, Swiss Federal Institute of Technology in Lausanne, Switzerland, 1998.

[PAH 99] PAHUD D., FROMENTIN A., HUBBUCH M., Heat exchanger pile system of the Dock Midfield at the Zürich Airport. Detailed simulation and optimisation of the installation, Final report, Swiss Federal Office of Energy, Publication number 195325, available at www.bfe.admin.ch/dokumentation/energieforschung, 1999.

[PAH 07a] PAHUD D., HUBBUCH M., Mesures et optimisation de l'installation avec pieux énergétiques du Dock Midfield de l'aéroport de Zürich, Final Report, Federal Office of Energy, Publication number 270095, available at www.bfe.admin.ch/dokumentation/energieforschung, 2007.

[PAH 07b] PAHUD D., PILESIM2: simulation tool for heating and cooling systems with energy piles or multiple borehole heat exchangers, User manual, ISAAC – DACD – SUPSI, Switzerland, 2007.

List of Authors

Ghassan Anis AKROUCH
Texas A&M University
USA

Nahed ALSHERIF
University of Colorado
Boulder
USA

Tony AMIS
GI Energy Ltd
United Kingdom

Jean-Baptiste BERNARD
ECOME Ingénierie
France

Peter BOURNE-WEBB
Instituto Superior Tecníco
Lisbon
Portugal

G. Allen BOWERS
VirginiaTech
USA

Jean-Louis BRIAUD
Texas A&M University
USA

Sébastien BURLON
LGCgE
University of Lille
France

Charles J. R. COCCIA
University of Colorado
Boulder
USA

Alice DI DONNA
Swiss Federal Institute of Technology
Lausanne
Switzerland

Fabrice DUPRAY
Swiss Federal Institute of Technology
Lausanne
Switzerland

Wolf FRIEDEMANN
Ed Züblin AG
Stuttgart
Germany

Julien HABERT
LGCgE
University of Lille
France

Ghazi HASSEN
Ecole des Ponts ParisTech
France

Lyesse LALOUI
Swiss Federal Institute of Technology
Lausanne
Switzerland

John S. MCCARTNEY
University of Colorado
Boulder
USA

Thomas MIMOUNI
Swiss Federal Institute of Technology
Lausanne
Switzerland

Hussein MROUEH
LGCgE
University of Lille
France

C. Guney OLGUN
VirginiaTech
USA

Daniel PAHUD
University of Applied Sciences and
Arts of Southern Switzerland
Switzerland

Jean-Michel PEREIRA
Ecole des Ponts ParisTech
France

Norbert PRALLE
Ed Züblin AG
Germany

Marcelo SANCHEZ
Texas A&M University
USA

Melissa A. STEWART
University of Colorado
Boulder
USA

Maria E. SURYATRIYASTUTI
LGCgE
University of Lille
France

Anh Minh TANG
Ecole des Ponts ParisTech
France

Veli Matti UOTINEN
Ruukki
Finland

Nico VON DER HUDE
Bilfinger Berger AG
Germany

Bernhard WIDERIN
Enercret GmBH
Austria

Neda YAVARI
Ecole des Ponts ParisTech
France

Index

K, L

M

N

O

P

Q, R

S

T

U

V, W, Y

Printed and bound by CPI Group (UK) Ltd, Croydon, CR0 4YY

13/09/2026

14961076-0005